Math Mammoth
Grade 5 Review Workbook
A Complete Workbook with Lessons and Problems

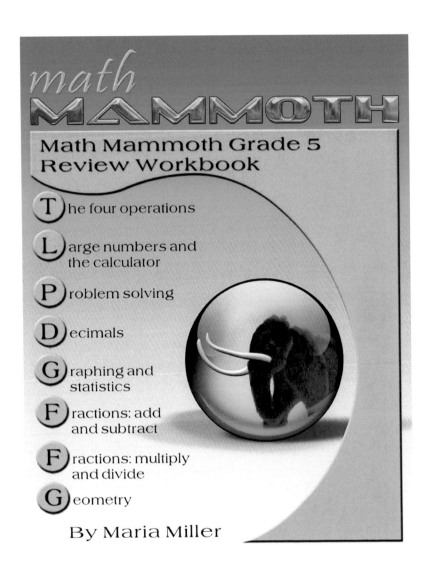

By Maria Miller

Copyright 2016 Maria Miller.
ISBN 978-1530619030

EDITION 3/2016

All rights reserved. No part of this book may be reproduced or transmitted in any form or by any means, electronic or mechanical, or by any information storage and retrieval system, without permission in writing from the author.

Copying permission: Permission IS granted for the teacher to reproduce this material to be used with students, not commercial resale, by virtue of the purchase of this book. In other words, the teacher MAY make copies of the pages to be used with students. Permission is given to make electronic copies of the material for back-up purposes only.

Math Mammoth Grade 5 Review Workbook
Contents

Introduction	5
The Four Operations Review	7
The Four Operations Test	11
Large Numbers and the Calculator Review	13
Large Numbers and the Calculator Test	17
Mixed Review 1	19
Mixed Review 2	21
Problem Solving Review	23
Problem Solving Test	27
Mixed Review 3	29
Mixed Review 4	31
Decimals Review	33
Decimals Test	39
Mixed Review 5	43
Mixed Review 6	45
Graphing and Statistics Review	47
Graphing and Statistics Test	49
Mixed Review 7	51
Mixed Review 8	54
Fractions: Add and Subtract Review	57
Fractions: Add and Subtract Test	59
Mixed Review 9	61
Mixed Review 10	64
Fractions: Multiply and Divide Review	67
Fractions: Multiply and Divide Test	71
Mixed Review 11	73
Mixed Review 12	76
Geometry Review	79
Geometry Test	83
Mixed Review 13	85

Mixed Review 14 .... **88**
End-of-the-Year Test **91**

Answers 105

Introduction

Math Mammoth Grade 5 Review Workbook is intended to give students a thorough review of fifth grade math, following the main areas of Common Core Standards for grade 5 mathematics. The book has both topical as well as mixed (spiral) review worksheets, and includes both topical tests and a comprehensive end-of-the-year test. The tests can also be used as review worksheets, instead of tests.

You can use this workbook for various purposes: for summer math practice, to keep the child from forgetting math skills during other break times, to prepare students who are going into sixth grade, or to give fifth grade students extra practice during the school year.

The topics reviewed in this workbook are:

- the four operations
- large numbers and the calculator
- problem solving
- decimals
- graphing and statistics
- fractions: add and subtract
- fractions: multiply and divide
- geometry

In addition to the topical reviews and tests, the workbook also contains many cumulative (spiral) review pages.

The content for these is taken from the *Math Mammoth Grade 5 Complete Curriculum*, so this workbook works especially well to prepare students for grade 6 in Math Mammoth. However, the content follows a typical study for grade 5, so this workbook can be used no matter which math curriculum you follow.

Please note this book does not contain lessons or instruction for the topics. It is not intended for initial teaching. It also will not work if the student needs to completely re-study these topics (the student has not learned the topics at all). For that purpose, please consider *Math Mammoth Grade 5 Complete Curriculum*, which contains all the necessary instruction and lessons.

I wish you success with teaching math!

Maria Miller, the author

The Four Operations Review

1. Solve (without a calculator).

   a. 7,587 ÷ 27

   b. 2,829 ÷ 41

   c. 249 × 382

2. Solve 83,493 − y = 21,390.

3. Solve in the right order. You can enclose the operation to be done first in a "bubble" or a "cloud."

a. 5 × (3 + 8) = _____	**b.** 20 + 240 ÷ 8 + 90 = _____
c. 100 − 2 × 5 × 7 = _____	**d.** 70 − 2 × (2 + 5) = _____

4. Divide mentally, and solve in the right order.

a. $\dfrac{3636}{6}$ =	**b.** $\dfrac{3608}{4}$ =	**c.** $\dfrac{4050}{5}$ =
d. 42 + $\dfrac{255}{5}$ =		**e.** $\dfrac{4{,}804}{2+2}$ =

5. Find a number to fit in the box so the equation is true.

| a. $25 = 7 + \Box \times 2$ | b. $72 \div 8 = (6 - 3) \times \Box$ | c. $(4 + \Box) \div 3 = 2 + 2$ |

6. Write an expression _or_ an equation to match each written sentence. You do not have to solve.

a. The difference of x and 9	b. The sum of y and 3 and 8 equals 28.
c. The quotient of 60 and b is equal to 12.	d. The product of 8, x and y

7. Which expression matches the problem? Also, solve the problem.

Three girls divided equally the cost of buying four sandwiches for $3.75 each. How much did each girl pay?	(1) $3 \times \$3.75 - 4$	(2) $3 \times \$3.75 \div 4$
	(3) $\$3.75 \div 4 \times 3$	(4) $4 \times \$3.75 \div 3$

8. Write a <u>single</u> expression (number sentence) for the problems, and solve.

a. Bonnie and Ben bought an umbrella for $12 and boots for $17, and divided the cost equally. How much did each pay?

b. Henry bought five jugs of milk for $4.50 each. In the end, the grocer gave him $2 off his bill. What did Henry pay?

9. Draw a bar model to represent the equations. Then solve them.

a. R ÷ 4 = 544

b. 4 × R = 300

10. Mark an "x" if the number is divisible by 2, 3, 5, 6, or 9.

Divisible by	2	3	5	6	9
534					
123					

Divisible by	2	3	5	6	9
1,605					
2,999					

11. Factor the following numbers to their prime factors.

a. 21
/ \

b. 12
/ \

c. 38
/ \

d. 75
/ \

e. 124
/ \

f. 89
/ \

The Four Operations Test

1. Solve (without a calculator).

   a. $1{,}456 \div 26$

   b. $18{,}755 \div 31$

   c. 391×475

2. Solve: $Y - 8{,}687 = 19{,}764$

3. Solve in the right order.

a. $2 \times (80 - 8) = $ _____	**b.** $100 - 240 \div (8 + 2) = $ _____

4. Divide mentally.

a. $\dfrac{8{,}109}{9} = $	**b.** $\dfrac{1{,}244}{4} = $	**c.** $\dfrac{4{,}045}{4 + 1} = $

5. Find a number to fit in the box so the equation is true.

a. $42 = (\ \square\ - 10) \times 2$	**b.** $48 \times 10 = \square \times 6$

6. Write an expression *or* an equation to match each written sentence. You do not have to solve anything.

a. The product of *s* and 11	b. The quotient of 48 and *b* is equal to 8.

7. Write a <u>single</u> expression (number sentence) for the problem, and solve.

Mia bought 5 pairs of socks for $2.50 each, and paid with a $20 bill. What was her change?

8. Draw a bar model to represent each equation and solve the equation.

a. $5 \times Y = 600$	b. $Z \div 3 = 140$

9. Is 991 divisible by 3?

   Why or why not?

10. Factor the following numbers to their prime factors.

a. 16 /\	b. 34 /\	c. 80 /\

Large Numbers and the Calculator Review

1. Write the numbers.

    a. 560   70 thousand   9 million

    b. 60 million   5 hundred   7 thousand   4 tens

    c. 50 billion   50   50 thousand

    d. 98 million   431 billion   940

2. What is the *place* and the *value* of the underlined digit?

a. 405,2<u>2</u>9,020	b. 97,02<u>4</u>,003,245
Place: _____	Place: _____
Value: _____	Value: _____
c. 2<u>3</u>0,560,079,000	d. 4,<u>5</u>89,211,000
Place: _____	Place: _____
Value: _____	Value: _____

3. Round these numbers to the nearest thousand, nearest ten thousand, nearest hundred thousand, and nearest million.

number	69,066	14,506,439	389,970,453	12,976,895,322
to the nearest 1,000				
to the nearest 10,000				
to the nearest 100,000				
to the nearest million				

4. Read the powers aloud. Then write out the repeated multiplications, and solve.

a. $8^2 =$	d. $1^5 =$
b. $4^3 =$	e. $100^2 =$
c. $10^3 =$	f. $2^5 =$

5. Write using exponents, and solve.

a. $3 \times 3 \times 3 =$	e. $10 \times 10 \times 10 \times 10 \times 10 =$
b. $7 \times 7 =$	f. $2 \times 2 \times 2 \times 2 \times 2 \times 2 =$
c. five squared =	g. five cubed =
d. ten cubed =	h. ten to the sixth power =

6. Calculate the products mentally.

a. $200 \times 110{,}000$	b. $30 \times 200 \times 600$
c. $50{,}000 \times 200{,}000$	d. 6×10^3
e. 8×10^5	f. 2×10^8
g. 21×10^4	h. 829×10^6

7. Add $7{,}890{,}483 + 32{,}930 + 155{,}670$ without using a calculator.

8. Complete the addition path.

56,700 → add 20,000 → ☐ → add a 100,000 → ☐

add ↓ a million

☐ ← add 10 million ← ☐ ← add 2 billion ← ☐

9. First estimate the answer. Then calculate the exact answer and the error of estimation using a calculator.

a. 6,808 + 493,420	b. 3,703 × 52,380
My estimation: _____	My estimation: _____
Exact answer: _____	Exact answer: _____
Error of estimation: _____	Error of estimation: _____

10. In 2007, there were about 308,397 new vehicles sold each week in the United States. The average price of a new vehicle was $28,800. Calculate the total amount of money spent on buying vehicles in one week. (Source: Nada.org)

11. On average, the money spent to educate one pupil for one year in public school in the USA is $9,138. There are about 4,050,000 fifth-graders in the United States. *Estimate* using rounded numbers about how much money is spent yearly to educate them in public schools.

Who am I? 	*Who am I?* 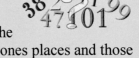
My digit in the ten thousands place is double the digit in the tens place.	I have the same digit in the thousands, millions, and ones places and those add up to 9.
The digit in the tens and hundreds places add up to 7.	The tens digit is four times the ten thousands digit, and both are even numbers more than zero.
I have 2 in the millions place.	The hundred thousands digit and hundreds digit add up to 1, the former being less than the latter.
In the hundred thousands place, I have an even digit that is less than 4 but more than 0.	
All the rest of my digits are zeros.	
All total, my digits add up to 17.	

Large Numbers and the Calculator Test

The calculator is not allowed in the first six problems of the test.

1. Write the numbers.

    a. 70 million 6 thousand 324

    b. 4 billion 32 thousand

    c. 98 billion 89 million 98

2. What is the *value* of the underlined digit?

a. 410,2<u>9</u>3,004	**b.** 408,0<u>3</u>7,443,000	**c.** <u>4</u>,395,490,493
Value: _____	Value: _____	Value: _____

3. Round these numbers as indicated.

number	183,602	355,079,933	29,928,900
to the nearest 1,000			
to the nearest 10,000			
to the nearest 100,000			
to the nearest million			

4. Solve.

a. $9^2 =$ _____	**b.** $10^3 =$ _____	**c.** $3^3 =$ _____

5. Write using exponents, and solve.

a. six squared =	**b.** two to the fifth power =

6. Calculate the products mentally.

a. 40 × 900,000	**b.** 600 × 200 × 500
c. 7×10^4	**d.** 48×10^6

7. Complete the math path. (Calculator usage is optional.)

| 4 million | →subtract 700 thousand→ | | →add 12 million→ | |

add ↓ 3 billion

| | ←subtract 20 thousand← | | ←add 8 hundred← | |

8. First, estimate the answer. Then calculate the exact answer and the error of estimation using a calculator.

a. 209,800 − 4,730	**b.** 2,543 × 5,187
Estimation: _____	Estimation: _____
Exact answer: _____	Exact answer: _____
Error of estimation: _____	Error of estimation: _____

c. 56,493,836 + 345,399 + 7,089,400

Estimation: _____

Exact answer: _____   Error of estimation: _____

9. In October 2008, the US congress approved a $700 billion "bailout plan" to aid the failing banks. There were about 305,000,000 persons in the USA. If the cost of this bailout had been divided evenly between all the people of the U.S., how much would a family of four in the USA have had to pay for it? Give your answer rounded to the nearest hundred dollars.

18

Mixed Review 1

1. Find the missing number and write if it is the **minuend**, **subtrahend** or the **difference**.

a. 76 − _____ = 11	b. _____ − 39 = 18	c. 48 − 29 = _____
_____	_____	_____

2. Write an equation to match the bar model, and solve it.

| x | 9,380 | 3,928 |

←————— 93,450 —————→

3. Solve.

a. $10 \times 2 \times 2 \times 3 \times 100 \times 7$	b. $400 \times 3,000 \times 110$	c. $500 \times 200 \times 300 \times 10$
d. 10^4	e. 3^3	f. $10^6 \times 7$

4. Which power of ten is equal to one million?

5. There are 365 days in a year. How many hours are there in a year?

   _____

   Estimate: _____

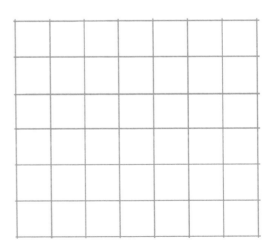

6. Divide. Use the space on the left for building a multiplication table of the divisor. Lastly check.

	a. $43 \overline{)5\ 5\ 0\ 4}$	$\times\ 4\ 3$
	b. $82 \overline{)7\ 7\ 9\ 0}$	$\times\ 8\ 2$

7. Factor the following numbers to their prime factors.

a. 28 /\	**b.** 98 /\	**c.** 66 /\
d. 17 /\	**e.** 51 /\	**f.** 53 /\

Mixed Review 2

1. Place parentheses into these equations to make them true.

| a. $90 + 70 + 80 \times 2 = 390$ | b. $378 = 6 \times 8 + 13 \times 3$ | c. $90 \times 4 = 180 - 60 \times 3$ |

2. Draw a bar model to illustrate the equations. Then solve the equations.

a. $4x + 120 = 200$

b. $25 + 3x = 52$

3. Divide. Use the space on the left for building a multiplication table of the divisor. Lastly, check.

$2 \times 15 = 30$

a. $15 \overline{)9450}$

$\times\ 1\ 5$

b. $14 \overline{)4508}$

$\times\ 1\ 4$

4. Solve the word problems.

   a. Jim earned a total of $1,920 dollars in four weeks.
   How much did he earn in one week?

   b. Joe entered his sled and dogs in 11 races last year.
   The races were all held on the same 136-mile race course.
   How many miles total did Joe and his dogs race last year?

5. Which expression(s) match the problem? Also, solve the problem.

Greg bought four flashlights for $9 each, and paid with $50. What was his change?	(1) $50 − $9 + $9 + $9 + $9 (2) $50 − ($9 − $9 − $9 − $9) (3) $50 − ($9 + $9 + $9 + $9)	(4) 4 × $9 − $50 (5) $50 − 4 × $9 (6) $50 + 4 − $9

6. First, estimate the answer to the multiplication problem. Then multiply.

a. Estimate: _____

$$\begin{array}{r} 173 \\ \times\ 35 \\ \hline \end{array}$$

b. Estimate: _____

$$\begin{array}{r} 269 \\ \times\ 537 \\ \hline \end{array}$$

c. Estimate: _____

$$\begin{array}{r} 892 \\ \times\ 340 \\ \hline \end{array}$$

Problem Solving Review

1. Solve in the right order. You can do these mentally, if you remember the "trick!"

a. $\dfrac{633}{2+1} =$	b. $\dfrac{1{,}000 + 555}{7 - 2} =$

2. Draw a bar model to illustrate the equations. Then solve the equations.

a. $2x + 72 = 164$	b. $420 + 5x = 1{,}080$

3. Joey has already saved some money. This coming Saturday he will earn $25, and then the next Tuesday he will earn $10. After that, he will have just enough money to buy a used bicycle for $109. So how much had Joey saved before Saturday?

4. Solve the problems. Mark the given numbers in the model. Mark what is not known with "?".

a. Eva and Jane canned a total of 137 quarts of peaches.
Jane canned 45 more quarts than Eva.
How many quarts did Eva can?

b. Jay and Joe each built a raft. Jay's raft was only 2/3 as long as Joe's. Joe's raft was 3 feet longer than Jay's.
How long was each raft?

c. Bill was driving from home to his grandparents' farm. After 112 miles he stopped to get gas. After another 35 miles he was at the half-way point.

How many miles is it from his home to his grandparents' farm?

d. A pack of 5 light bulbs costs $7.50.
What would 8 light bulbs cost?

5. Dad gave $175 to his two sons, Austin and Brandon, so that Brandon got $37 more than Austin. How much did Austin get?

6. A $364 video camera is on sale with 1/4 off of the normal price. What is the discounted price?

7. Three-fourths of the chess club members are 18 years of age or older. If 69 members are 18 or older, how many members does the club have?

Problem Solving Test

1. Write an equation to match each balance. Then solve what x stands for.
   Remember to write 2x to mean 2 x's in the same pan, and 3x to mean x, x, and x in the same pan.

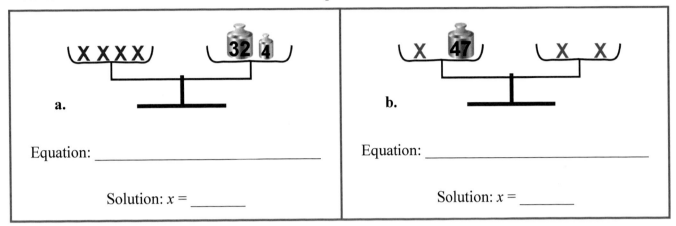

a.

Equation: _____

Solution: $x = $ _____

b.

Equation: _____

Solution: $x = $ _____

2. Write an equation for each bar model. Then, solve for x.

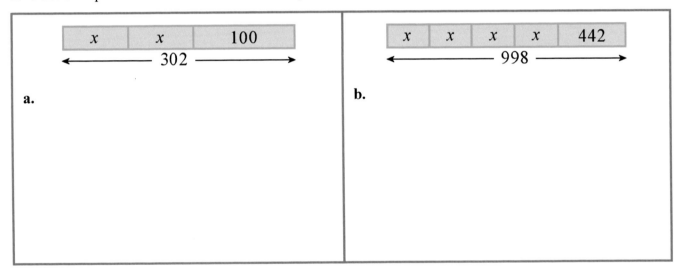

a.

b.

3. A cell phone that costs $48 is on sale with 1/6 off of the normal price.
   How much would *three* discounted phones cost?

4. Two sisters divided 250 smooth beach rocks so that the elder sister had 32 rocks more than the younger sister. How many rocks did the younger sister get?

5. Five kilograms of potatoes cost $7.50. Henry bought 2 kg.

   a. How much do 2 kg of potatoes cost?

   b. What was Henry's change from $10?

6. A high-quality hard drive costs three times as much as a low-quality one. Buying the two together would cost $820. How much does the low-quality hard drive cost?

7. Matthew is 3/8 as tall as his dad. If Matthew is 66 cm tall, then how tall is his dad?

Mixed Review 3

1. Draw a bar model where the total is 547, and the three parts are 119, 38, and x. Lastly solve for x.

2. The washer uses about 14 gallons of water for a load of laundry. If you run the washer three times a week, how much water do you use in a year?

3. Write either an expression _or_ an equation to match each written sentence.

a. The difference of 16 and 7 is 9.
b. The sum of 3, 9, and y is 20.

4. Find the number missing from the box to make the equation true.

a. $42 = (7 + \Box) \times 2$	**b.** $480 \div 8 = 10 \times 5 + \Box$	**c.** $4 + \Box = (200 - 50) \div 2$

5. Which of the following calculations can be used to check the division $458 \div 7 = 65$ R3 ?

   a. $3 \times 65 \times 7$ **b.** $65 + 7 \times 3$ **c.** $7 \times 65 + 3$ **d.** $(7 + 65) \times 3$

6. Determine if the two expressions would have the same value, without actually calculating.

a. $3{,}289 - 144 - 276$	**b.** $613 - (325 - 249)$	**c.** 5×636
$3{,}289 - (144 + 276)$	$613 - 325 - 249$	$2 \times 636 + 2 \times 636$

7. Factor the following numbers to their prime factors.

a. 64 /\	b. 60 /\	c. 85 /\

8. Divide. Use the space on the left for building a multiplication table of the divisor. Lastly check.

	79)8 9 2 7	× 7 9

9. Fill in the missing parts.

a. $2 \times 10^4 =$ _____	b. $712 \times 10^3 =$ _____	c. $55 \times 10^6 =$ _____
d. $6 \times 10^\square = 6{,}000$	e. $18 \times 10^\square = 180{,}000{,}000$	f. $69 \times \square^\square = 69{,}000{,}000$

10. First estimate the answer by using rounded numbers. Next, calculate the exact answer with a calculator. Then, find the error of estimation with a calculator.

a. $15{,}278 \times 3{,}892$ (round to thousands)	b. $19{,}945{,}020 - 6{,}320{,}653$ (round to millions)
My estimation: _____	My estimation: _____
Exact answer: _____	Exact answer: _____
Error of estimation: _____	Error of estimation: _____

Mixed Review 4

1. Divide. Use the space on the right for building a multiplication table of the divisor. Then check.

2 × 21 = 42	21) 8 1 6 9	× 2 1

2. Solve in the right order. First, you can enclose the operation to be done in a "bubble" or a "cloud."

a. 94 + 12 × 5 ÷ 2 = _____	b. (22 − 9) × 2 + 58 = _____
c. 43 + (55 + 5) ÷ 5 = _____	d. 700 − 30 × (3 + 4) = _____

3. Solve mentally.

a. 43 − 17 = _____	b. 54 − 19 + 12 = _____	c. 1,200 − _____ = 750
71 − 43 = _____	85 − 25 + 75 = _____	2,000 − 800 − _____ = 600

4. Write the numbers.

   a. 78 billion   38   16 thousand

   b. 844 billion   12 million   704

5. Round these numbers to the nearest thousand, nearest ten thousand, nearest hundred thousand, and nearest million.

number	32,274,302	64,321,973	388,491,562	2,506,811,739
to the nearest 1,000				
to the nearest 10,000				
to the nearest 100,000				
to the nearest million				

6. Complete the addition path using mental math.

| 43,199,000 | add 10,000 → | | add a million → | |

add ↓ 100 thousand

| | ← add 10 million | | ← add a thousand | |

7. Write an expression to match each written sentence.

a. The product of 5 and 6 is added to 50.	**b.** The difference of 9 and 6 is subtracted from 10.

8. Write a single expression using numbers and operations for the problem, not just the answer!

A teacher bought 21 notebooks for $2 each, 20 rulers for $1.50 each, and chalk for $12. What was the total cost?

9. Add.

a. 521,607,090 + 4,293,991,092

b. 77,630,087 + 884,000,299 + 84,926,571

10. Estimate first, using mental math. Then find the exact answer and the error of your estimation using a calculator.

a. 2,933 × 213

My estimation: _____

Exact answer: _____

Error of estimation: _____

b. 152 × 89 × 7,932

My estimation: _____

Exact answer: _____

Error of estimation: _____

Decimals Review

1. Color parts to show the decimals.

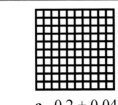		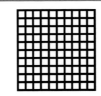	
a. 0.2 + 0.04	**b.** 0.09 + 0.05	**c.** seven hundredths	**d.** 0.6

2. Write in expanded form.

   a. 0.495

   b. 2.67

3. Write the decimals indicated by the arrows.

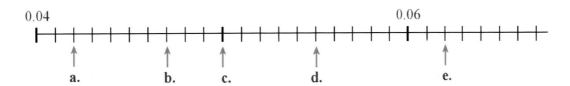

a. _____ b. _____ c. _____ d. _____ e. _____

4. Compare using $<$, $=$, and $>$.

a. 0.25 ☐ 0.215	**b.** 0.3 ☐ 0.19	**c.** 4.033 ☐ 4.33	**d.** 0.65 ☐ $\frac{1}{2}$	**e.** 0.065 ☐ 0.2

5. Write as decimals.

   a. $\frac{3}{100} =$ _____ **b.** $\frac{48}{1000} =$ _____ **c.** $1\frac{209}{1000} =$ _____ **d.** $3\frac{39}{100} =$ _____

6. Write as fractions or mixed numbers.

   a. 1.3   **b.** 2.15

   c. 0.008   **d.** 0.038

7. Round the numbers to the nearest one, to the nearest tenth, and to the nearest hundredth.

rounded to...	nearest one	nearest tenth	nearest hundredth
4.608			
3.109			
2.299			
0.048			

8. Add and subtract.

a. $0.3 + 0.005 =$ _____ $0.03 + 0.5 =$ _____	b. $0.9 - 0.7 =$ _____ $0.9 - 0.07 =$ _____
c. $0.008 + 0.9 + 5 =$ _____ $0.9 + 0.8 + 0.17 =$ _____	d. $2.5 - 1.02 =$ _____ $7.8 - 0.9 - 0.04 =$ _____

9. Complete the addition sentences.

a. $0.21 +$ _____ $= 1$	b. $0.004 +$ _____ $= 1$	c. $4.391 +$ _____ $= 5$

10. **a.** Find the number that is 5 hundredths and 7 tenths *more* than 3.194.

   b. Find the number that is 3 thousandths and 8 tenths *less* than 0.902.

11. Five children divided $25 equally, and then each one bought ice cream for $2.05.
    a. *Estimate* how much each child has left now.
    b. Choose an expression that matches the problem.
    c. Find the exact amount that each child has left now.

   $25 - \$2.05 \div 5$
   $25 \div 5 - \$2.05$
   $25 - 5 \times \$2.05$

12. Solve.

a. $0.4 \times 8 =$ _____	c. $20 \times 0.5 =$ _____	e. $0.9 \times 0.2 =$ _____
b. $6 \times 0.009 =$ _____	d. $100 \times 0.3 =$ _____	f. $0.06 \times 0.3 =$ _____

13. Divide.

a. $0.35 \div 5 =$ _____	c. $0.4 \div 10 =$ _____	e. $0.38 \div 10 =$ _____
b. $4.5 \div 9 =$ _____	d. $5 \div 100 =$ _____	f. $7 \div 1000 =$ _____

14. Find the missing factors.

a. $0.8 \times$ _____ $= 0.40$	c. $7 \times$ _____ $= 3.5$	e. $0.9 \times$ _____ $= 7.2$
b. $8 \times$ _____ $= 0.064$	d. $0.6 \times$ _____ $= 0.024$	f. $9 \times$ _____ $= 0.81$

15. Multiply and divide using powers of ten.

a. $0.07 \times 10^2 =$ _____	b. $3{,}300 \div 10^4 =$ _____
$10^5 \times 1.08 =$ _____	$239.8 \div 10^3 =$ _____

16. Use decimal multiplication to find these amounts.

a. 7/10 of 5 kg	b. 6/100 of 1.2 meters	c. 35/100 of 2 liters

17. Multiply and divide. Use the grid.

   a. 2.3×0.79 b. $2.485 \div 7$ c. $17.0 \div 20$

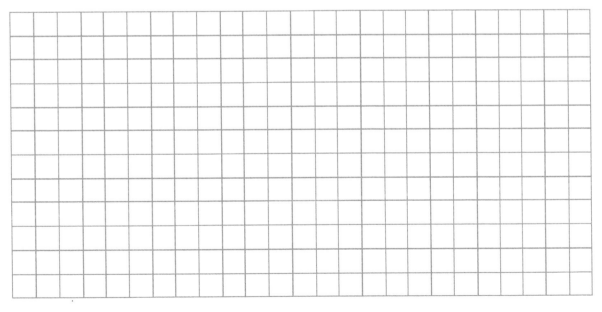

18. Is the answer to 0.4 × 0.7 less than or more than 0.7?

    Explain in your own words why that is so.

19. A bicycle that costs $126 is discounted by 2/10 of its price.
    Find the discounted price.

20. Multiply both the dividend and the divisor by 10, repeatedly, until you get a *whole-number divisor*.
    Then, divide using long division. If the division is not even, round the answer to two decimals.

a. 152.8 ÷ 0.4	b. 2.776 ÷ 0.08
c. 180 ÷ 1.1	d. 2 ÷ 7

21. Convert.

a. 0.9 m = _____ cm	b. 0.6 L = _____ ml	c. 2.2 kg = _____ g
45 cm = _____ m	5,694 ml = _____ L	390 g = _____ kg
1.5 km = _____ m	0.09 L = _____ ml	0.02 kg = _____ g

22. Convert.

a. 6 ft 11 in. = _____ in.	b. 2 gal = _____ C	c. 78 oz = _____ lb _____ oz
3 lb 11 oz = _____ oz	5 qt = _____ pt	39 in = _____ ft _____ in
3 C = _____ oz	54 oz = _____ C _____ oz	102 in = _____ ft _____ in

23. Convert. Use a calculator, but only in this problem!

a. 2.65 mi. = _____ ft	b. 3,800 ft = _____ mi	c. 4.54 lb = _____ oz
10.9 mi = _____ yd	3,500 yd = _____ mi	10.2 ft = _____ in

24. Twenty-six kilograms of strawberries are packaged evenly into five boxes.

   a. How much does each box weigh?

   b. If the strawberries cost $3 per kilogram, how much does one box cost?

25. Edward earns $11.75 per hour. Find his earnings in a 38-hour week. Then figure out what he takes home after paying 1/5 of it in taxes.

26. Two pitchers hold a total of 3.65 liters. The smaller pitcher holds 0.55 L less than the larger one. Find the individual volumes of the two pitchers.

Decimals Test

As this test is quite long, feel free to administer it in two parts.

1. Write the decimals indicated by the arrows.

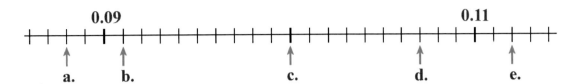

a. _____ b. _____ c. _____ d. _____ e. _____

2. Complete the addition sentences.

| a. 1.3 + _____ = 7 | b. 0.76 + _____ = 1 | c. 3.65 + _____ = 4 | d. 0.18 + _____ = 0.2 |

3. Write as decimals.

| a. $\frac{21}{100}$ = | b. $\frac{46}{1000}$ = | c. $3\frac{7}{100}$ = | d. $20\frac{2}{10}$ = |

4. Write as fractions or mixed numbers.

   a. 0.6   b. 0.82   c. 1.208   d. 0.093

5. Compare.

   a. 0.05 ☐ 0.2   b. 0.43 ☐ 0.045   c. 2.05 ☐ 2.051   d. 0.438 ☐ $\frac{1}{2}$

6. Round the numbers to the nearest one, nearest tenth, and nearest hundredth.

rounded to...	nearest one	nearest tenth	nearest hundredth	rounded to...	nearest one	nearest tenth	nearest hundredth
8.816				0.398			
1.495				9.035			

7. Solve.

| a. 0.4 × 7 = _____ | c. 20 × 0.05 = _____ | e. 0.2 × 1,000 = _____ |
| b. 7 × 0.09 = _____ | d. 100 × 0.09 = _____ | f. 0.8 × 0.8 = _____ |

8. Divide.

a. $0.24 \div 6 =$ _____	c. $2 \div 100 =$ _____	e. $0.43 \div 10 =$ _____
b. $0.081 \div 9 =$ _____	d. $0.8 \div 10 =$ _____	f. $7 \div 1000 =$ _____

9. Multiply and divide using powers of ten.

a. $0.05 \times 10^4 =$ _____	c. $3.5 \div 10^2 =$ _____
b. $10^5 \times 7.8 =$ _____	d. $13{,}200 \div 10^4 =$ _____

10. Find the number that is 1 tenth and 2 thousandths more than 1.109.

11. **a.** Estimate the answer to 0.6×21.8.

   b. Now find the exact answer to 0.6×21.8.

12. Divide using long division:

   a. $7.836 \div 6$

   b. $21 \div 4$

13. Is the answer to 0.9 × 0.8 more or less than 0.8?

    Explain in your own words why that is so.

    Is it more or less than 0.9?

14. Teresa packed 7 kg of blueberries equally into four boxes.
    How much does each box weigh?

15. Convert.

a. 0.7 m = _____ cm	b. 2,650 ml = _____ L	c. 5.16 kg = _____ g
3.2 km = _____ m	0.9 L = _____ ml	400 g = _____ kg

16. Convert.

a. 8 ft 10 in = _____ in	b. 2 gal 3 C = _____ C	c. 81 oz = _____ lb _____ oz
183 in = _____ ft _____ in	45 oz = _____ C _____ oz	165 oz = _____ lb _____ oz

17. Samuel bought a 0.9-liter box of juice and two cans of juice, 350 ml each.
    What is the total volume of the juice he bought?

18. Mary bought a 2-kg bag of tomatoes for $4.48.
    Then, Mary sold 250 g of the tomatoes to her friend.

    a. How much does half a kilogram of tomatoes cost?

    b. How much did she charge her friend?

19. A DVD that normally costs $19.95 is discounted.
    The new price is 2/5 off of the normal price.
    Find how much *two* copies of the discounted DVD cost.

20. Arnold weighed the apples they had left in the root cellar in the Spring.
    He gave half of the apples to his neighbor, and divided the rest equally into four boxes.
    Each box weighed 0.47 kg.

    a. In the model, label the parts that equal 0.47 kg.

    b. In the model, label the total weight
    of all the apples with "???"

    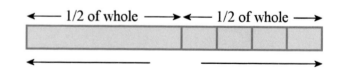

    c. Find the total weight of the remaining apples that were stored in the cellar.

Mixed Review 5

1. Divide mentally.

| a. $\dfrac{7490}{7} =$ | b. $\dfrac{5030}{2} =$ | c. $\dfrac{5406}{6} =$ |

2. Solve the equations.

a. $83{,}493 - y = 21{,}390$

$y = $ _____

b. $20 \times s = 6{,}340$

$s = $ _____

3. Solve in the right order. You can enclose the operation to be done first in a "bubble" or a "cloud."

a. $5 + (6 + 9) \div 3 = $ _____	b. $20 \times 12 \div 3 - 50 = $ _____
c. $100 - 36 \div 6 \times 7 = $ _____	d. $(88 - 3 \times 5) \times 2 = $ _____

4. Write these numbers.

a. 24 + 600 thousand + 15 billion =

b. 42 million + 17 + 80 thousand =

5. Find all the factors of the given numbers.

a. 42	b. 64
Check 1 2 3 4 5 6 7 8 9 10	Check 1 2 3 4 5 6 7 8 9 10
factors: _____	factors: _____

6. Solve.

a. Fifty song downloads cost $5.50.
How much would 20 downloads cost?

b. Mr. Doe paid 1/6 of his $870 salary in taxes, and $140 as a loan payment. How much of his salary was left?
Mark the information from the problem on the diagram.
Mark with "?" what the problem asks for.

c. Tommy owns 450 stamps, which is one-fourth of the amount that Henry owns. How many does Henry own?

d. Dad bought two hammers. One cost $28 more than the other, and the total cost was $64. How much did the cheaper hammer cost?

Who am I?

"I'm lurking between 200 and 300...
My ones digit is double my hundreds digit.
I'm divisible by 8 and by 11."

Mixed Review 6

1. Jake earned $125 and his sister earned 4/5 as much. How much did Jake and his sister earn together?

   Mark the information in the bar model, and solve.

   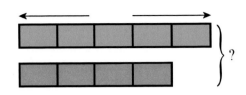

2. Mom is buying a thermometer, and the store has two kinds. The pricier one costs $10.40, and the cheaper just 3/4 as much.

   How much more does the expensive thermometer cost than the cheaper one?

3. Helen has 120 marbles and Julie has 2/5 as many. How many more marbles does Helen have than Julie?

4. Ann is an English teacher. She has 150 students in her English classes this year, and 6/50 of them were not in her classes last year.

   a. How many new students does she have?

   b. Out of the new students, 1/3 have never studied English before. How many of the new students have studied English before?

5. Divide. Use the space on the right for building a multiplication table of the divisor. Then check.

2 × 37 = 74	37) 6 7 3 4	× 3 7

6. Solve for the unknown N or M.

a. 4 × M = 200	b. M ÷ 6 = 8	c. 4,500 ÷ M = 50
d. 7 × N = 56,000	e. N ÷ 30 = 700	f. 48,000 ÷ N = 600

7. Write an expression to match each written sentence.

a. The quotient of 350 and x equals 5.	b. The difference of 15 and 6 is added to 8.

8. Find a number to fit in the box so the equation is true.

a. 36 = (☐ + 9) × 3	b. 7 × 7 = 4 × ☐ + 5	c. 19 = (84 ÷ ☐) − 2

9. Round these numbers to the nearest thousand, nearest ten thousand, nearest hundred thousand, and nearest million.

number	97,302	25,096,199	709,383,121	89,534,890,066
to the nearest 1,000				
to the nearest 10,000				
to the nearest 100,000				
to the nearest million				

Graphing and Statistics Review

1. Plot the points from the "number rule" on the coordinate grid. Fill in the rest of the table first, using the rule given.

   The rule is: $y = 9 - x$.

x	0	1	2	3	4
y					

x	5	6	7	8	9
y					

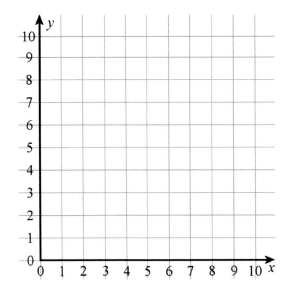

2. Find the mean and mode of this data set to the nearest hundredth: 5, 9, 13, 12, 16, 10, 19, 11, 10.

3. **a.** Estimate what the amount of tractors might have been in the year 2010.

   b. During which decade did the amount of tractors rise the quickest?

   What was the *approximate* amount of increase in tractors during that decade?

   c. Describe the trend in the amount of tractors between 1970 and 1995.

   d. About how many-fold was the increase in tractors between 1930 and 1960?

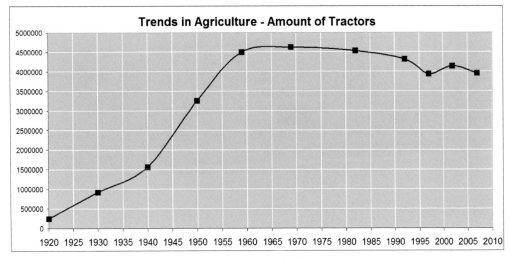

Source: Census of Agriculture

4. A department store was tracking the sales of many items, including umbrellas.

   a. In 2007, which months were the sales less than 40 umbrellas? How about in 2008?

   b. Find the month with the greatest difference between 2007 and 2008 sales.

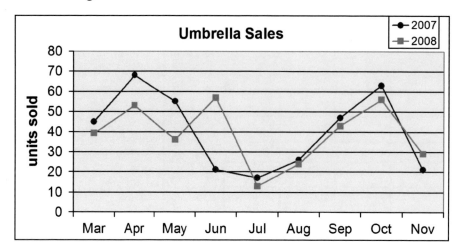

5. Four hundred eight students were asked about how many hours they had slept the previous night. The results are summarized in the table below:

Hours of Sleep	Frequency
6	2
7	15
8	56
9	148
10	137
11	40
12	10
total	408

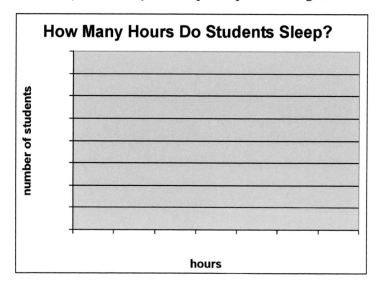

   a. Draw a bar graph. Note you need to choose the scaling on the vertical axis.

   b. Find the mode.

   c. Which of the following could be a part of the *original* data set? (Hint: Imagine the original data set. Think: What was asked? What did the students answer?)

   2, 15, 56, 148, 137, 40, 10, 2, 15, ...

   6, 10, 8, 8, 9, 7, 11, 10, 9, 10, 11,...

   d. Which of the following is the average for this data set?

   9.4 hours   8.5 hours   11.1 hours   10.7 hours

Graphing and Statistics Test

1. Plot the point (9, 5) on the grid. Then, plot the point that is two units down and four units to the left from that point. What are its coordinates?

2. Plot the points from the "number rule" on the coordinate grid. Fill in the rest of the table first, using the rule.

    The rule is: $y = 2x - 1$.

x	1	2	3	4
y				

x	5	6	7	8
y				

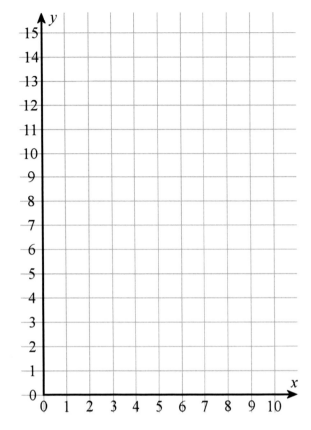

3. A store kept track of how many cell phones they sold each day of the week ("units sold"). On a certain day they started a promotion with 3/10 off of the normal price.

    a. Add the number labels for the vertical axis next to the tick marks (the scaling).

    b. Plot the remaining points and finish the line graph.

    c. Which day did the promotion most likely start?

Day	Units sold
Mo	17
Tu	14
Wd	15
Th	21
Fr	19
Sa	23
Mo	15
Tu	34
Wd	40
Th	37
Fr	33
Sa	41

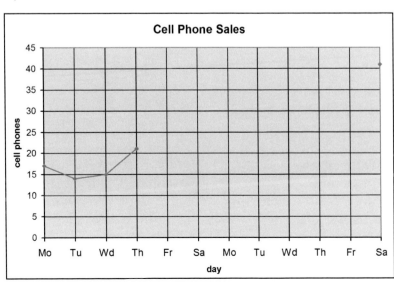

4. Mary asked 20 people in a club how old they were (in years). Here is her data:
   10, 9, 10, 12, 15, 8, 9, 10, 11, 13, 11, 10, 9, 12, 9, 13, 11, 10, 15, 14
   (Each number is the response from one person.)

   a. Fill in the frequency table. Make four categories. Draw a bar graph.

Age	frequency

   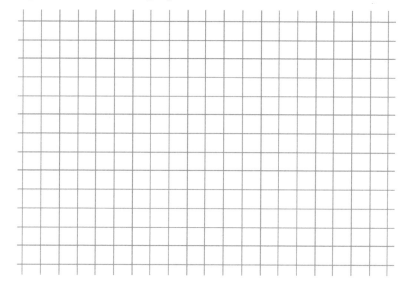

   b. What is the mode of this data set?

   c. Find the average.

5. The chart shows Alice's science test scores for five different tests.

Alice's test scores	
Test 1	76
Test 2	66
Test 3	74
Test 4	81
Test 5	88

   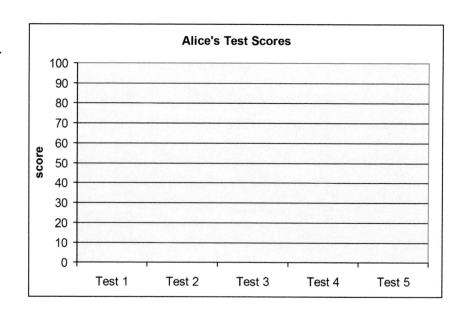

   a. Draw a line graph.

   b. Calculate the average.

   c. Plot the average on the line graph.

   d. If the test where Alice scored the worst was dismissed and not taken into account, what would Alice's average score be?

Mixed Review 7

1. **a.** Write a number that is 5 thousandths, 2 tenths, and 8 hundredths more than 1.004.

   b. Write a number that is 3 thousandths and 3 tenths less than 3.411.

2. Figure out what was done in each step - either addition or subtraction!

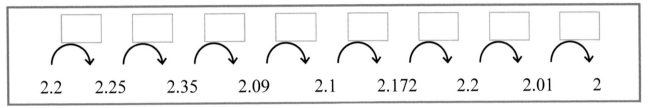

   2.2   2.25   2.35   2.09   2.1   2.172   2.2   2.01   2

3. Multiply mentally.

a.	b.	c.	d.
2 × 0.06 = _____	0.4 × 0.7 = _____	100 × 0.12 = _____	1.1 × 0.9 = _____
2 × 0.6 = _____	5 × 0.007 = _____	0.5 × 0.03 = _____	1000 × 0.05 = _____

4. **a.** Estimate the total cost in dollars.

   b. Find the total.

   c. Find the error of estimation.

beans	$4.35
milk	$2.99
dog food	$11.38
broccoli	$2.14
chicken	$7.64

   Estimate:

5. Factor the following numbers to their prime factors.

a. 48 /\	b. 71 /\	c. 93 /\

6. Find the value of x.

    | 1.46 | 1.46 | 1.46 | x |

    ←——— 10 ———→

7. Calculate.

   a. $2 \times 10^4 =$ _____   **b.** $9 \times 10^6 =$ _____   **c.** $17 \times 10^3 =$ _____

8. Write a division equation where the quotient is 210, the divisor is 52, and the dividend is unknown. Use a letter for the unknown. Then find the value of the unknown.

9. Mark the numbers given in the problem in the bar model. Mark what is asked with "?". Then solve the problem.

   *Mary and Luisa bought a gift together for $46.
   Mary spent $6 more on it than Luisa.
   How many dollars did each spend?*

   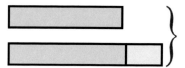

10. With the money John earned from his summer job, he paid his phone service for two months ($48 per month), spent $120 for a bike, and still had half of his money left. How much did he earn?

11. Each story in a tall apartment building is 235 cm high. Estimate the total height of the 12-story building, in meters.

12. Divide. Check your answer by multiplying.

a. 38) 3 9 5 2	× 3 8 _____	b. 17) 2 6 8 . 6	× _____

13. Divide.

a. Calculate 56 ÷ 9 to two decimal digits.) _____	b. Change the problem 5.175 ÷ 0.5 so that you get a *whole-number divisor*. Then, divide.) _____

14. First, estimate the answer by using rounded numbers. Then calculate the exact answer with a calculator. Lastly, find the error of estimation with a calculator.

a. 127,285 + 84,662 (round to thousands) My estimation: _____ _____ Exact answer: _____ Error of estimation: _____	b. 12,705,143 − 6,460,788 (round to millions) My estimation: _____ _____ Exact answer: _____ Error of estimation: _____

Mixed Review 8

1. Round the numbers to the nearest unit (one), to the nearest tenth, and to the nearest hundredth.

Round this to the nearest →	unit (one)	tenth	hundredth
4.925			
6.469			

Round this to the nearest →	unit (one)	tenth	hundredth
5.992			
9.809			

2. Jake worked for 56 days on a farm, and Ed worked for 14 days less. How many days did the two boys work together?

3. Add using mental math

a. 0.3 + 0.07 = _____	b. 0.19 + 0.002 = _____	c. 0.028 + 0.3 = _____
d. 1.05 + 0.4 = _____	e. 0.49 + 0.56 = _____	f. 0.006 + 0.5 = _____

4. Multiply both the dividend and the divisor by 10, repeatedly, until you get a *whole-number divisor*. Then, divide using long division.

a. $0.927 \div 0.3$

Check:

b. $0.646 \div 0.08$

Check:

5. Convert the measuring units.

a. 0.5 m = _____ cm	b. 4.2 L = _____ mL	c. 800 g = _____ kg
0.06 m = _____ cm	400 mL = _____ L	4,550 m = _____ km
2.2 km = _____ m	5,400 g = _____ kg	2.88 kg = _____ g

6. Jerry bought three packages of AA batteries and six packages of AAA batteries. The total was $17.04. One package of the AA batteries cost $1.88. What does one package of AAA batteries cost?

7. Divide in two ways: first by indicating a remainder, then by long division. Give your answers to two decimal digits.

a. 31 ÷ 6 = _____ R _____	b. 43 ÷ 4 = _____ R _____
) _____ Check:	) _____ Check:

8. First, estimate the answer. Then, multiply. Do not forget the decimal points.

a. 2.09 × 11.5	b. 73 × 2.14	c. 7.1 × 3.02
Estimate: _____	Estimate: _____	Estimate: _____
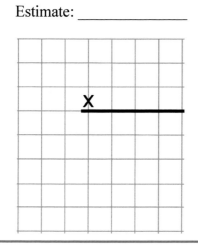	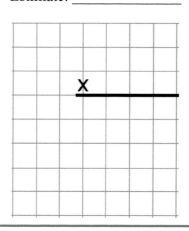	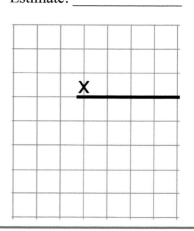

9. Alex bought seven packets of cucumber seeds for $13.23 total. He also bought seven flower plants that were originally $3.20 each but the price was reduced by 4/10.

   a. What did one packet of seeds cost?

   b. How much did one flower plant cost?

   c. What was the total cost?

10. Multiply and divide.

a. 10 × 0.07 = _____	b. 100 × 0.63 = _____	c. 1000 × 0.029 = _____
d. 0.8 ÷ 10 = _____	e. 4.5 ÷ 100 = _____	f. 76 ÷ 1000 = _____

Fractions: Add and Subtract Review

1. Write as fractions. Think of the shortcut.

| a. $9\frac{1}{2}$ | b. $5\frac{6}{11}$ | c. $8\frac{2}{7}$ | d. $5\frac{6}{100}$ |

2. Write as mixed numbers.

| a. $\frac{41}{10}$ | b. $\frac{19}{3}$ | c. $\frac{28}{9}$ | d. $\frac{32}{12}$ |

3. For the division problem 23 ÷ 6 = 3 R5, write a corresponding problem where a fraction is changed into a mixed number.

4. Subtract. Regroup if necessary. Check that your answer is reasonable.

| a. $9\frac{4}{8}$ $-3\frac{7}{8}$ | b. $12\frac{3}{20}$ $-5\frac{11}{20}$ | c. $10\frac{3}{5}$ $-5\frac{1}{3}$ |

5. Add and subtract. Check that your answer is reasonable.

| a. $\frac{5}{7} + \frac{1}{3}$ | b. $\frac{3}{10} + \frac{1}{3}$ |
| c. $3\frac{2}{7} - 1\frac{6}{7}$ | d. $2\frac{4}{5} + 3\frac{1}{4}$ |

6. Compare the fractions, and write <, >, or = in the box.

a. $\frac{1}{2} \square \frac{3}{5}$	b. $\frac{3}{11} \square \frac{1}{3}$	c. $\frac{7}{10} \square \frac{70}{100}$	d. $\frac{1}{4} \square \frac{28}{100}$
e. $\frac{2}{3} \square \frac{8}{9}$	f. $\frac{1}{4} \square \frac{2}{15}$	g. $\frac{21}{16} \square \frac{25}{16}$	h. $\frac{5}{11} \square \frac{1}{2}$

7. Betty uses 3 1/8 feet of material to make one shirt. She has one piece that is 5 1/2 feet and another piece that is 4 1/4 feet. She made one shirt from *each* piece of material.
How much material does she have left now?

8. Of a piece of land, 32/100 is planted in wheat, 42/100 is planted in barley, 2/10 is planted in oats, and the rest is resting.
What part (fraction) of the land is resting?

9. Which is a better deal: 1/5 off of a book that costs $35, or 2/11 off of a book that costs $33?

Would the situation change if both deals involved a book that costs $50? Explain.

Fractions: Add and Subtract Test

1. Write as mixed numbers.

   a. $\dfrac{26}{3}$

   b. $\dfrac{45}{7}$

   c. $\dfrac{34}{5}$

2. Add or subtract.

a. $7\dfrac{6}{8}$ $+\ 2\dfrac{5}{8}$	b. $6\dfrac{1}{5}$ $-\ 3\dfrac{4}{5}$	c. $4\dfrac{6}{11}$ $+\ 9\dfrac{9}{11}$ $+\ 2\dfrac{4}{11}$

3. Find the missing fractions or mixed numbers.

a. $2\dfrac{3}{7} +\ \underline{} = 5\dfrac{1}{7}$	b. $2\dfrac{5}{9} + 4\dfrac{6}{9} + \underline{} = 10$	c. $7\dfrac{2}{15} - \underline{} = 2\dfrac{8}{15}$

4. Mark the fractions on the number line. $\dfrac{2}{3},\ \dfrac{5}{6},\ \dfrac{7}{12},\ \dfrac{3}{4},\ \dfrac{11}{12}$

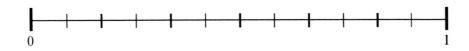

5. If you can find an equivalent fraction, write it. If you cannot, cross out the whole problem.

a. $\dfrac{3}{7} = \dfrac{\ }{21}$	b. $\dfrac{4}{3} = \dfrac{\ }{18}$	c. $\dfrac{5}{6} = \dfrac{\ }{11}$	d. $\dfrac{2}{5} = \dfrac{8}{\ }$	e. $\dfrac{5}{6} = \dfrac{15}{\ }$

6. Compare the fractions, and write <, >, or = in the box.

a. $\dfrac{7}{4}\ \square\ \dfrac{5}{3}$	b. $\dfrac{5}{11}\ \square\ \dfrac{1}{2}$	c. $\dfrac{7}{10}\ \square\ \dfrac{69}{100}$	d. $\dfrac{3}{4}\ \square\ \dfrac{75}{100}$	e. $\dfrac{8}{7}\ \square\ \dfrac{7}{9}$

7. Draw something in the picture and explain how we can add 1/3 and 2/5.

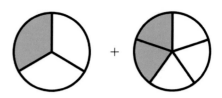

8. Add and subtract.

a. $\dfrac{2}{3} + \dfrac{3}{4}$	b. $\dfrac{5}{6} - \dfrac{2}{3}$
c. $3\dfrac{1}{7} - \dfrac{1}{2}$	d. $6\dfrac{7}{8} + 3\dfrac{1}{5}$

9. Write the fractions in order starting from the smallest.

$\dfrac{4}{7}, \dfrac{5}{9}, \dfrac{7}{5}, \dfrac{1}{2}$

10. Write a fraction addition with a sum (answer) of 1 where one fraction has a denominator of 4, and the other has a denominator of 7.

11. Measure the sides of the triangle in inches. Find its perimeter.

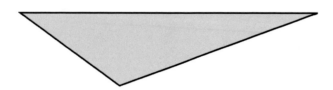

Mixed Review 9

1. In what place is the underlined digit? What is its value?

a. 452,9<u>1</u>2,980	b. <u>6</u>,219,455,221
Place: _____	Place: _____
Value: _____	Value: _____

2. How many seconds are in an hour?

   How many seconds are in a day?

3. The Hewitts family homeschools all but 12 weeks of the year, five days a week, about five hours a day. How many hours of schoolwork do they do in a year?

4. Solve by multiplying in columns. Estimate first.

   2.11 × 6.8

   Estimate: _____

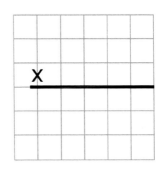

5. Multiply and divide.

a. 0.34 ÷ 10 = _____	b. 100 × 0.098 = _____	c. 19 ÷ 10^3 = _____
2.1 ÷ 100 = _____	1,000 × 46.7 = _____	10^4 × 0.03 = _____

6. Convert between the measuring units.

a. 5,070 g = _____ kg	b. 0.6 L = _____ ml	c. 0.06 km = _____ m
2.5 kg = _____ g	10,500 ml = _____ L	2,600 m = _____ km

7. First, multiply the divisor and dividend by 10, 100, or 1000 so that the divisor becomes a whole number. Then divide.

a. $82.50 \div 0.06$	b. $48.302 \div 0.2$

8. One cup of plain yogurt costs $2.40, a cup of strawberry yogurt costs $0.15 less than plain yogurt, and a cup of plum yogurt costs $0.30 more than plain yogurt. What is the total cost if you buy one cup of each kind of yogurt?

9. The price of Shirt A is $6.29. It is one-third of the price of Shirt B. Find the price of Shirt B.

10. Fill in Rachel's reasoning for solving $1{,}000 \times 0.007$.

Multiplying as if there was no decimal point, I get $1{,}000 \times$ _____. That equals _____.

Then, since my answer has to have thousandths, it needs _____ decimal digits.

So, the final answer is _____.

11. **a.** Make a histogram out of the data in the frequency table on the right.

Height in cm	Number of people
120...129	4
130...139	10
140...149	41
150...159	82

Height in cm	Number of people
160...169	95
170...179	61
180...189	39
190...199	6

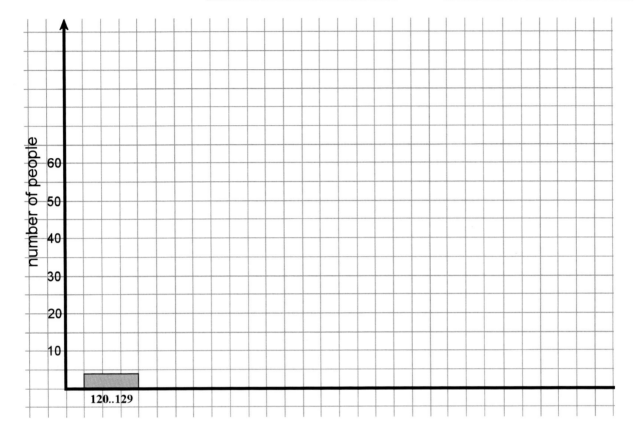

b. How many people were short (less than 140 cm tall)?

c. How many were tall (180 cm or taller)?

d. Most adults are 160 cm tall or taller. Use this fact to guess (estimate) how many children and how many adults were in this group.

e. Could this data come from

- a group of elementary school children?

- a group of people who were at the swimming pool at 5 pm on a certain Tuesday?

- a group of elderly women in an old people's home?

Explain your reasoning.

Mixed Review 10

1. First *estimate* the answer to each multiplication. Then multiply to find out the exact answer.

a. 290 × 277	**b.** 525 × 416	**c.** 897 × 186
Estimate: _____	Estimate: _____	Estimate: _____
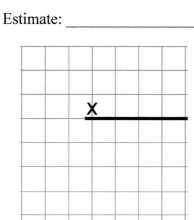	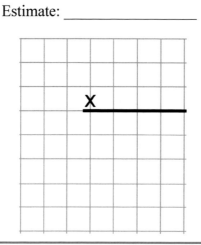	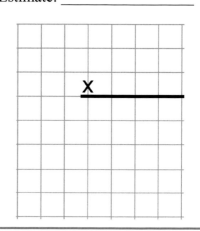

2. While jogging, Rebecca saw a big snake on the path 250 m before the end of the 2.4-km jogging track. She got so scared that she turned back on the track and jogged back to the beginning of the track. Find the total distance that she jogged on the track.

3. Angi and Rebekkah split their total earnings of $100 so that Angi got $10 more than Rebekkah. How much did each one get?

4. Find the missing factor.

a. 10 × _____ = 4.0	**b.** 5 × _____ = 6.0	**c.** _____ × 0.11 = 3.3
d. _____ × 0.3 = 0.06	**e.** 2 × _____ × 1.2 = 48	**f.** 3 × _____ × 0.5 = 6

5. Solve the equations.

a. $y - 0.57 = 1.1$	b. $7.319 + z = 9$

6. Calculate the average (the mean) of the data set. Do not use a calculator.

   21, 19, 25, 22, 13, 15, 24, 12, 11

7. Write in expanded form.

   a. 0.908

   b. 543.2

8. Divide. Mental math will work!

a. $0.8 \div 2 =$ _____	b. $0.36 \div 6 =$ _____	c. $0.25 \div 0.05 =$ _____
d. $0.16 \div 4 =$ _____	e. $0.54 \div 0.06 =$ _____	f. $1 \div 0.05 =$ _____

9. A group of 37 medical students traveled through ten states to view new technology in some progressive hospitals. They had to share the expense of $99,000 for the trip. What was each student's share of the expenses? Round your answer to the nearest dollar.

10. Ashley bought a 1/2-gallon carton of milk, and used 2 cups of it for baking. How many *cups* of milk are left?

11. Ava is 4 ft 8 in. tall and Eva is 61 inches tall. Who is taller? How many inches taller?

12. Juan is mailing 36 CDs that weigh 6 ounces each. What is the total weight of the CDs, in pounds and ounces?

13. The following numbers describe the distance in kilometers that 16 employees of a small company drive to work.   15 7 22 6 16 25 31 45 7 11 9 19 25 4 15 18

   a. Fill in the frequency table below. Make four or five categories. Then draw a histogram.

   b. Find the average number of kilometers the employees drive to work.

   c. Find the mode of the data set.

distance	frequency

Fractions: Multiply and Divide Review

1. Simplify the fractions. Draw pie pictures of the process.

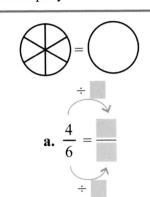 a. $\dfrac{4}{6} = \dfrac{}{}$	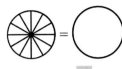 b. $\dfrac{9}{12} = \dfrac{}{}$	c. $\dfrac{24}{30} =$	d. $3\dfrac{15}{35} =$
		e. $\dfrac{56}{49} =$	f. $\dfrac{12}{100} =$
		g. $\dfrac{45}{27} =$	h. $2\dfrac{72}{84} =$

2. Draw a picture to illustrate these calculations, and solve.

a. $3 \times 1\dfrac{1}{3}$	b. $2 \times \dfrac{5}{6}$

3. Multiply.

a. $7 \times \dfrac{2}{5}$	b. $\dfrac{2}{7} \times \dfrac{5}{6}$
c. $4\dfrac{3}{10} \times 4$	d. $1\dfrac{1}{6} \times 5\dfrac{2}{3}$

4. Simplify before you multiply.

a. $\dfrac{7}{14} \times \dfrac{3}{12}$

b. $\dfrac{5}{24} \times \dfrac{12}{30}$

5. Figure out the side lengths of the colored rectangle from the picture. Then multiply the side lengths to find its area. <u>Check that the area you get by multiplying is the same as what you can see</u> from the picture.

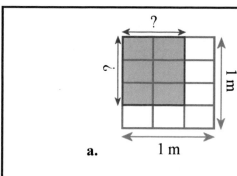

a.

Side lengths: ⬜/⬜ m and ⬜/⬜ m

Area: ⬜/⬜ m × ⬜/⬜ m =

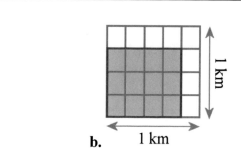

b.

Side lengths: ⬜/⬜ km and ⬜/⬜ km

Area: ⬜/⬜ km × ⬜/⬜ km =

6. Shade a rectangle inside the square so that its area can be found by the fraction multiplication.

a. $\dfrac{5}{6}$ m × $\dfrac{1}{2}$ m = ⬜/⬜ m²

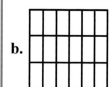

b. $\dfrac{2}{3}$ in. × $\dfrac{1}{6}$ in. = ⬜/⬜ in²

7. Mary jogs 3/4 of a mile each day, five days a week. Calculate how many miles she jogs in a week.

8. Sally made a rectangular blueberry pie and cut it into 20 equal pieces. The next morning, 12/20 of it was left. Then, the dog got on the table and gobbled up 2/3 of what was left!

   a. How many pieces are left now?

   b. What fraction of the pie is left now?

9. Draw a picture to illustrate these calculations, and solve.

a. $1 \div 3$	b. $\frac{1}{2} \div 3$

10. Divide.

a. $2 \div \frac{1}{3}$	b. $4 \div \frac{1}{4}$	c. $\frac{1}{2} \div 5$
d. $\frac{1}{7} \div 3$	e. $9 \div 4$	f. $\frac{1}{8} \div 2$
g. $6 \div 9$	h. $\frac{6}{10} \div 3$	i. $\frac{3}{4} \div 3$

11. Solve.

a. A string that is 7 inches long is cut into four equal pieces. How long are the pieces?

b. Find four-fifths of the fraction 1/3.

c. Five people are sharing equally 11 lb of almonds. How many pounds will each get?

d. Chain costs $24 per meter, and you bought 3/4 of a meter. What was the cost?

e. There were 112 contestants, and 3/8 of them were women. How many were not?

12. Is the result of multiplication more, less, or equal to the original number? You do not have to calculate anything. Compare and write <, >, or = in the box.

| a. $\frac{8}{9} \times 7$ ☐ 7 | b. $2\frac{1}{11} \times 57$ ☐ 57 | c. $\frac{7}{7} \times 13$ ☐ 13 |

13. One-third of a cake was decorated with chocolates, one-fourth with sprinkles, and the rest with strawberry frosting. What *part* was decorated with strawberry frosting?

14. A loaf of bread was cut into 30 slices. After a day, 5/6 of it was left. Then the family ate 1/5 of the *remaining* bread. How many slices are left now?

15. You only have 3/4 cup of walnuts in the cupboard, so you decide to make only 3/4 of the recipe. How much of each ingredient do you need?

Brownies

3 cups sweetened carob chips
8 tablespoons olive oil
2 eggs
1/2 cup honey
1 teaspoon vanilla
3/4 cup whole wheat flour
3/4 teaspoon baking powder
1 cup walnuts or other nuts

16. **a.** Determine which sheet of paper has the greater area:
(1) a 6 ½ in. by 8½ in. sheet or (2) a 5¾ in. by 9 in. sheet.

b. How much greater is the area of the larger sheet than the area of the smaller sheet, in square inches?

Fractions: Multiply and Divide Test

1. Write the simplification process.

   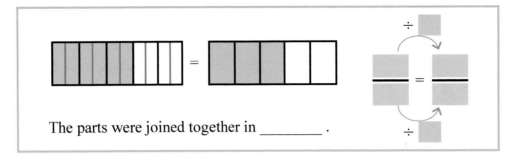

   The parts were joined together in _____ .

2. If possible, simplify the following fractions. Give your answer as a mixed number when possible.

a. $\frac{22}{6} =$	b. $\frac{28}{42} =$	c. $\frac{35}{32} =$

3. Julie needs 2/3 cups of butter for one batch of cookies. Find how much butter she would need to make five batches of cookies.

4. Draw a picture to illustrate $5 \times \frac{3}{4}$ and solve.

5. Is the following multiplication correct?

   If not, correct it.

   $\frac{3}{4} \times$ ◐ = ⊕

6. Multiply the fractions, and shade a picture to illustrate the multiplication. Simplify your answers.

   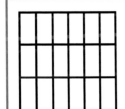 a. $\frac{2}{3} \times \frac{1}{6}$

    b. $\frac{4}{9} \times \frac{2}{3}$

7. Multiply. Give your answers in the lowest terms (simplified) and as a mixed number, if possible.

| a. $\dfrac{5}{12} \times \dfrac{2}{3}$ | b. $9 \times \dfrac{4}{5}$ |

8. Find the area of a square with 1 7/8-inch sides.

9. After supper, a family of four had 1/3 of a pizza left.
   The next day, three people shared the remaining pizza equally.
   What fractional part of the *original* pizza did each person get?

10. **a.** How many 1/3-lb servings can you get from 3 pounds of chicken?

    b. Write a division sentence to match this situation.

11. Solve.

| a. $\dfrac{1}{6} \div 3$ | b. $6 \div \dfrac{1}{8}$ | c. $\dfrac{9}{11} \div 3$ |

12. Draw a picture of some hearts, circles, and diamonds, so that 3/7 of the shapes are hearts, 2/7 of them are circles, and the rest are diamonds. What is the ratio of hearts to circles to diamonds?

13. A jar contains white and blue marbles in a ratio of 2:3. If there are 400 marbles in all, how many are white?

Mixed Review 11

1. Subtract.

a. $4\frac{2}{6} - 1\frac{5}{6} =$	b. $3\frac{2}{9} - 1\frac{7}{9} =$
c. $4\frac{2}{3} - 2\frac{1}{4} =$	d. $7\frac{1}{6} - 1\frac{3}{5} =$

2. Write the numbers in expanded form.

   a. 0.28

   b. 60.068

3. There were 780 people at the concert. One-third of them came with a discount ticket. Another 120 were seniors and came with a special low-price ticket. The rest paid a regular-priced ticket. How many people paid the normal price? *(You can draw a bar model to help.)*

4. Make a line graph of baby's weight.

Week	Weight	Weight in ounces
0	6 lb 14 oz	
1	6 lb 12 oz	
2	6 lb 14 oz	
3	7 lb	
4	7 lb 2 oz	
5	7 lb 4 oz	
6	7 lb 6 oz	
7	7 lb 7 oz	

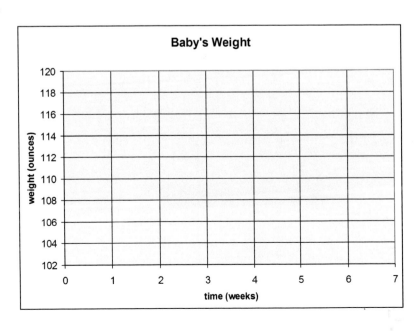

Plot the points from the "number rules" or number patterns on the coordinate grids.

5. **The rule for x-values:** start at 2, and add 1 each time.
   The rule for y-values: start at 0, and add 2 each time.

x	2					
y	0					

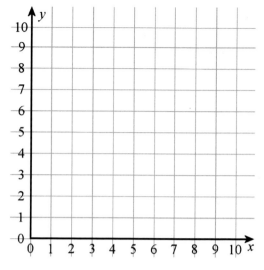

6. **The rule for x-values:** start at 1, and add 1 each time.
   The rule for y-values: start at 10, and subtract 2 each time.

x						
y						

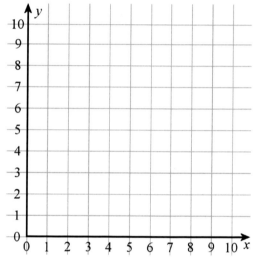

7. Use rounded numbers to estimate the answer.
   How many $0.58 cans can you get with $7?

8. Jack cut three 0.82-meter pieces from a 4-meter board.
   How long was the piece that was left?

9. Factor the following numbers to their prime factors.

a. 28 / \	b. 55 / \	c. 84 / \

10. This data are the responses of 20 fifth graders to the question, "What is your favorite pet?"

    dog, dog, hamster, dog, cat, cat, parrot, dog, horse, dog, cat, canary, goldfish, dog, cat, cat, dog, dog, canary, hamster

    a. Find the mode.

    b. If possible, calculate the mean.

11. Jenny buys two computer keyboards that had originally cost $15.60 but now they are 2/10 off of their normal price. Find Jenny's total bill.

12. Convert. One mile is 5,280 ft. Round your answers to whole feet.

    a. 0.6 mi = _____ ft

    b. 3.45 mi = _____ ft

13. Match the two problems (a) and (b) to the right expressions. Calculate. What does your answer mean?

    a. John has $170 to buy groceries for the week. First, John sets aside $23 to buy treats; then he divides the remaining money evenly for each day of the week.

    b. John has $170 to buy groceries for the week. John decides to use $23 per day for food, and to use whatever is left for treats.

    $170 − 7 × $23 $170 − $23 ÷ 7   ($170 − $23) ÷ 7

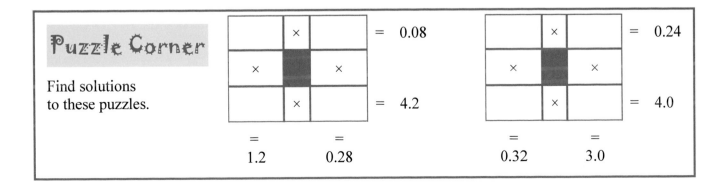

Puzzle Corner

Find solutions to these puzzles.

Mixed Review 12

1. Solve in the right order!

a.	b.	c.
$13 \times 4 + 18 = $ _____	$(2 + 60 \div 4) \times 3 = $ _____	$10 \times (9 + 18) \div 3 = $ _____
$4 + 8 \div 8 = $ _____	$2 + 30 \times (7 + 8) = $ _____	$5 \times (200 - 190 + 40) = $ _____

2. Joe bought 100 apples for $0.23 each. He divided them equally into ten small bags.

   a. What was the total cost for 100 apples?

   b. What was the value of each small bag of apples?

3. Compare the fractions.

   a. $\frac{2}{3}\ \square\ \frac{5}{8}$ **b.** $\frac{1}{4}\ \square\ \frac{4}{9}$ **c.** $\frac{5}{6}\ \square\ \frac{5}{7}$ **d.** $\frac{6}{8}\ \square\ \frac{3}{4}$

4. In what place is the underlined digit? What is its value?

a. 791,4<u>5</u>6,030	b. 2,09<u>4</u>,806,391
Place: _____	Place: _____
Value: _____	Value: _____

5. Make a line graph of this data for the Oak Bend Hospital.

Year	Babies Born
1950	225
1960	340
1970	460
1980	525
1990	580
2000	520
2010	490

6. Write an equation to match the bar model. Then, solve for x.

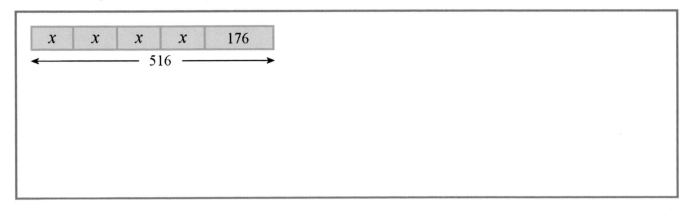

7. A hotel maintains two jogging paths in the woods. The shorter one is 1.2 km long and the other is four times as long. If you jog both paths, then what is the total distance you have jogged?

8. Shelly is going to buy eight pounds of oranges for $1.19 a pound, and six pounds of bananas for $0.88 a pound.

   a. Estimate the total cost to the nearest dollar.

   b. The cashier announces that Shelly will get 1/5 off of her bill for being a loyal customer. Now calculate what Shelly pays. Use the actual total in your calculation, not the rounded total.

9. Add and subtract.

a. $6\frac{6}{11} - 3\frac{2}{5}$	**b.** $6\frac{6}{7} + 1\frac{1}{2}$
c. $7\frac{9}{10} - 1\frac{1}{4}$	**d.** $3\frac{2}{5} + 2\frac{5}{6}$

10. These decimal divisions are not even. Round the answers to the nearest hundredth.

a. 3.377 ÷ 3	b. 22.91 ÷ 11	c. 62.6 ÷ 7

11. Divide in two ways: first by indicating a remainder, then by long division. Give your answers to two decimal digits.

a. 31 ÷ 6 = _____ R _____ Check:	b. 43 ÷ 4 = _____ R _____ Check:

Geometry Review

1. Measure all the angles of the triangles. Then classify the triangles.

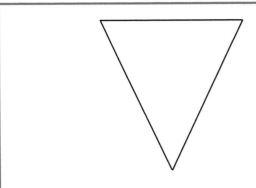

a. Angles: _____°, _____°, _____°

Acute, obtuse, or right?

Equilateral, isosceles, or scalene?

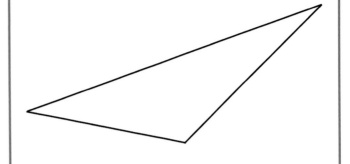

b. Angles: _____°, _____°, _____°

Acute, obtuse, or right?

Equilateral, isosceles, or scalene?

2. **a.** Draw an isosceles triangle with 50° base angles and a 7 cm base side (the side *between* the base angles).

   b. Measure the top angle.

   It is _____°.

   c. Find the perimeter of your triangle in millimeters.

3. Find the perimeter and area of this figure. All measurements are in inches.

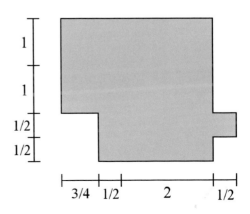

4. Name the different types of quadrilaterals.

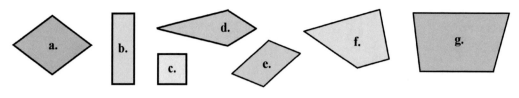

a.   b.

c.   d.

e.   f.

g.

5. Name the quadrilateral that...

   a. is a parallelogram and has four right angles

   b. is a parallelogram and has four sides of the same length

   c. has two parallel sides and two sides that are not parallel.

6. **a.** What is this shape called?

   b. Draw enough diagonals inside the shape to divide it into triangles.

   c. Number each of the triangles.

   d. Classify each triangle according to its sides (equilateral, isosceles, scalene) and according to its angles (acute, obtuse, right).

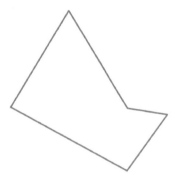

7. **a.** Draw an isosceles obtuse triangle.

   b. Draw a scalene acute triangle.

8. **a.** Draw a circle with its center at (2, 3) and a radius of 2 units. Use a compass.

   b. Draw another circle with its center at (6, 5) and a *diameter* of 8 units.

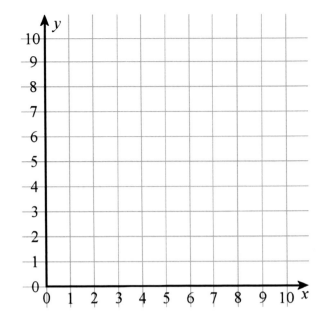

9. What is the height of this box, if its bottom dimensions are 2 cm × 4 cm and its volume is 32 cubic centimeters?

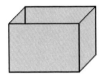

10. A gift box is 6 inches wide, 3 inches deep, and 2 inches tall. How many of these boxes do you need to have a total volume of 108 cubic inches?

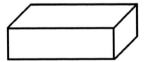

11. Find the volume of this rectangular prism, if...

    a. ...the side of each little cube is 1 inch.

    b. ....the side of each little cube is 2 inches.

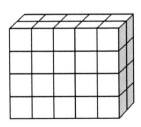

Puzzle Corner The area of the bottom of a cube is 16 cm². What is its volume?

Geometry Test

1. If the shape is a triangle, classify it by its sides *and* by its angles. If it is a quadrilateral, name it.

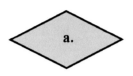   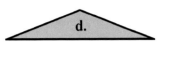 

   a. _____   b. _____

   c. _____

   d. _____

   e. _____

2. A certain quadrilateral has two pairs of parallel sides. Its sides measure 5 in, 5 in, 5 in, and 5 in.
   a. Is it also a kite?   **b.** A trapezoid?   **c.** A square?

3. Answer.

   a. Is a square also a kite?
   Why or why not?

   b. Is a rhombus also a trapezoid?
   Why or why not?

   c. If a quadrilateral has four right angles, can it be a kite?
   If yes, sketch an example.

   d. Could an equilateral triangle sometimes be a right triangle?
   If yes, sketch an example. If no, explain why not.

4. Which of these terms (perimeter, area, or volume) fits the situation, if you need to find out...

   a. ...how much fence is needed to go around a yard?

   b. ...how much water fits into a bottle?

   c. ...how big a carpet will cover the floor?

5. Draw an isosceles triangle with 30° base angles.
   You can choose the length of its sides.
   What is the measure of its top angle?

6. A square has a perimeter of 4 inches. What is its area?

7. The dimensions of this box are 2 ft by 1.5 ft by 1.5 ft.
   What is its volume?

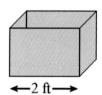

8. A book measures 15 cm × 30 cm × 1.5 cm.
   You make a stack of six books.

   a. What is the volume of one book?

   b. What is the volume of the stack?

9. **a.** Plot the points (0, 2), (0, 7), (4, 5), and connect them with line segments to form a triangle.

   Classify the triangle by its angles and sides. The triangle is

   _____ and

   b. Draw a circle with a center point of (4, 4) and a radius of 3 units.

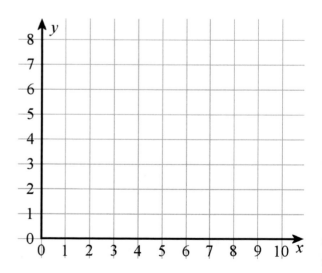

Mixed Review 13

1. Eric and Angela did yard work together. They earned $80 and split it so that Eric got $12 more than Angela. How much did each one get?

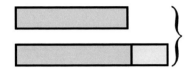

2. A bunch of five orchids costs $40, and a bunch of twenty daisies costs $40. Find the price difference between *one* orchid and *one* daisy.

3. Divide using long division. Check by multiplying.

a. 9,890 ÷ 46	b. 71.5 ÷ 65

4. Multiply and divide mentally.

a. 3 × 0.25 =	b. 8 × 0.08 =	c. 10 × 0.009 =
d. 0.9 × 8 =	e. 0.002 × 5 =	f. 2 × 0.3 × 7 =
g. 0.8 ÷ 4 =	h. 100 × 0.04 × 2 =	i. 7.2 ÷ 8 =
j. 0.8 ÷ 0.4 =	k. 2 ÷ 0.01 =	l. 0.056 ÷ 7 =

5. Solve.

a. $6 \times \dfrac{1}{5} =$	b. $\dfrac{1}{3} \times \dfrac{2}{7} =$	c. $\dfrac{6}{11} \times \dfrac{1}{8} =$
d. $\dfrac{10}{15} \times \dfrac{5}{6} =$	e. $3 \div \dfrac{1}{5} =$	f. $5 \div \dfrac{1}{3} =$
g. $\dfrac{1}{5} \div 2 =$	h. $\dfrac{1}{10} \div 3 =$	i. $7 \div 5 =$
j. $4 \div 9 =$	k. $40 \div 3 =$	l. $62 \div 9 =$

6. A company packs little jars of skin salve in boxes that are 6 inches tall. Each jar is 1 3/8 inches tall. How many of those jars can be stacked on top of each other in a box?

7. Solve for x.

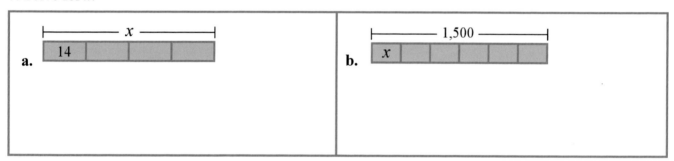

8. This morning one of her children is sick, so mom is making only 2/3 of her usual recipe. How much of each ingredient will she need? (*dl* stands for *deciliter*)

    What do you think she should do with the eggs?

> **Pancakes**
>
> 4 dl water
> 2 eggs
> 3 dl whole wheat flour
> (pinch of salt)
> 50 g butter for frying

9. Split the pieces further, and cross out some to show the subtractions. Solve.

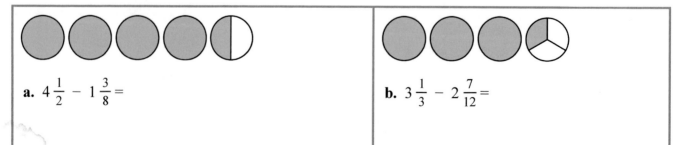

10. The ratio of green shirts to blue shirts is 3:5.

    a. What is the ratio of green shirts to all shirts?

    b. What is the ratio of blue shirts to all shirts?

    c. What is the ratio of blue shirts to green shirts?

11. The table shows how many adult and child visitors a small art museum had during one week.

    a. Calculate the total visitor counts.

    b. What was the difference in the total visitor count between the busiest day and the least busy day?

    c. Find the average number of adult visitors in a day. Give your answer to one decimal digit. Use your notebook for the long division.

    d. Find the average number of child visitors in a day. Give your answer to one decimal digit. Use your notebook for the long division.

Museum visitors			
Day	Adults	Children	*Total Visitors*
Monday	29	14	
Tuesday	23	10	
Wednesday	34	18	
Thursday	38	19	
Friday	35	19	
Saturday	57	25	
Sunday	63	31	
Totals			

    e. Make a double-bar graph of this data.

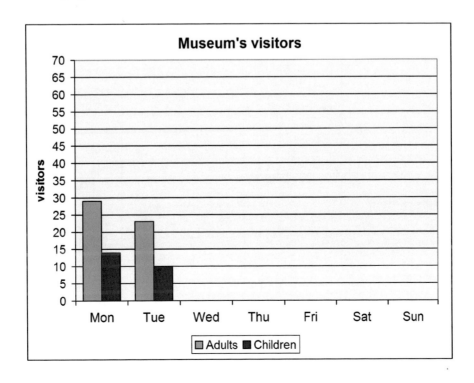

87

Mixed Review 14

1. Draw the hearts and diamonds as stated in the ratio in the box on the far right and fill in the fractions.

Picture or Diagram	As Fractions	As a Ratio
	___ of the shapes are hearts. ___ of the shapes are diamonds.	The ratio of hearts to diamonds is 1:5.

2. **a.** What is the ratio of circles to squares?

   b. What is the ratio of squares to all shapes?

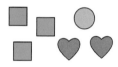

3. Jenny made gingersnaps, chocolate drops, and oatmeal cookies in the ratio of 1 : 3 : 2. She made 72 cookies in total. How many were oatmeal cookies?

4. Find the volume of a box that is 3 inches deep, 5 1/2 inches wide, and 2 inches tall.

5. Solve by multiplying in columns.

a. 21.7×3.9 Estimate: _____	**b.** 0.52×0.8 Estimate: _____	**c.** 141×5.22 Estimate: _____

88

6. Scott is a plumber, and each day he has to drive around town to the clients' homes.
   The following numbers show how many kilometers Scott drove on ten different workdays:

   128 68 73 163 93 102 68 85 90 45

   a. Find the average number of kilometers
      Scott drove per day.

   b. Based on the average, calculate *approximately*
      how many kilometers Scott would drive at work
      in a year's time. Assume that he works 40 weeks
      a year, 5 days a week.

7. Name the following quadrilaterals.

   a. _____

   b. _____

   c. _____   d. _____

8. Draw an isosceles right triangle
   with two 6-cm sides.

9. Multiply. Give your answers in the lowest terms (simplified) and as a mixed number, if possible.

   a. $\dfrac{6}{8} \times \dfrac{2}{9}$

   b. $\dfrac{9}{11} \times 2\dfrac{1}{3}$

10. A drinking glass measures 3/10 of a liter.
How many glasses full of water do you get from a 3-liter pitcher?

11. **a.** Fill in the table how much weight Greg gained during each year.

   b. At what ages did he gain weight the fastest?

   c. How can you see these fast growth periods on the graph?

AGE (yrs)	WEIGHT (kg)	Weight gain from previous year
0	3.3 kg	-
1	10.2 kg	6.9 kg
2	12.3 kg	2.1 kg
3	14.6 kg	
4	16.7 kg	
5	18.7 kg	
6	20.7 kg	
7	22.9 kg	
8	25.3 kg	
9	28.1 kg	

AGE (yrs)	WEIGHT (kg)	Weight gain from previous year
10	31.4 kg	
11	32.4 kg	
12	37.0 kg	
13	40.9 kg	
14	47.0 kg	
15	52.6 kg	
16	58.0 kg	
17	62.7 kg	
18	65.0 kg	

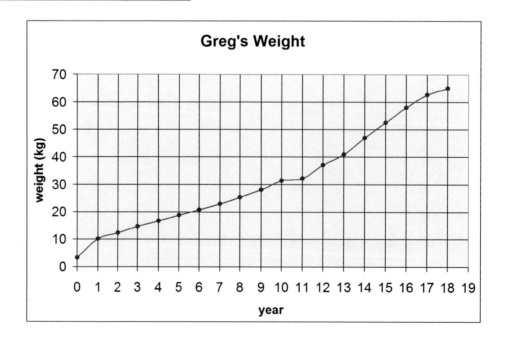

End-of-the-Year Test - Grade 5

This test is quite long, because it contains questions on all of the major topics covered in *Math Mammoth Grade 5 Complete Curriculum*. Its main purpose is to be a diagnostic test: to find out what the student knows and does not know. The questions are quite basic and don't involve especially difficult word problems.

Since the test is so long, I don't recommend that you have your child or student do it in one sitting. Break it into 3-5 parts and administer them on consecutive days, or perhaps in a morning/evening/morning/evening. Use your judgment.

A calculator is not allowed.

The test is evaluating the student's ability in the following content areas:

- the four operations with whole numbers
- the concept of an equation; solving simple equations
- divisibility and factoring
- place value and rounding with large numbers
- solving word problems, especially those that involve a fractional part of a quantity
- the concept of a decimal and decimal place value
- all four operations with decimals, to the hundredths
- coordinate grid, drawing a line graph, and finding the average
- fraction addition and subtraction
- equivalent fractions and simplifying fractions
- fraction multiplication
- division of fractions in special cases (a unit fraction divided by a whole number, and a whole number divided by a unit fraction)
- classifying triangles and quadrilaterals
- area and perimeter
- volume of rectangular prisms (boxes)

In order to continue with the *Math Mammoth Grade 6 Complete Worktext*, I recommend that the child gain a minimum score of 80% on this test, and that the teacher or parent review with him any content areas in which he may be weak. The exception to this rule is integers, because they will be reviewed in detail in 6th grade. Children scoring between 70% and 80% may also continue with grade 6, depending on the types of errors (careless errors or not remembering something, versus a lack of understanding). Again, use your judgment.

Instructions to the student:
Do not use a calculator. Answer each question in the space provided.

Instructions to the teacher:
My suggestion for points per item is as follows. The total is 171 points. A score of 137 points is 80%.

Question #	Max. points	Student score
The Four Operations		
1	2 points	
2	6 points	
3	2 points	
4	2 points	
5	2 points	
6	2 points	
7	3 points	
	subtotal	/ 19
Large Numbers		
8	2 points	
9	1 point	
10	1 point	
11	4 points	
	subtotal	/ 8
Problem Solving		
12	3 points	
13	3 points	
14	3 points	
15	3 points	
16	3 points	
17	3 points	
	subtotal	/ 18
Decimals		
18	4 points	
19	6 points	
20	3 points	
21	3 points	
22	3 points	
23	3 points	
24	9 points	
25	6 points	
26	9 points	
27	3 points	
28	3 points	
	subtotal	/52

Question #	Max. points	Student score
Graphs		
29	3 points	
30	2 points	
31	4 points	
	subtotal	/9
Fractions		
32	3 points	
33	4 points	
34	4 points	
35	2 points	
36	4 points	
37	2 points	
38	5 points	
39	3 points	
40	2 points	
41	4 points	
42	2 points	
43	2 points	
44	4 points	
	subtotal	/41
Geometry		
45	4 points	
46	4 points	
47	2 points	
48	3 points	
49	3 points	
50	3 points	
51	1 point	
52	4 points	
	subtotal	/24
	TOTAL	/171

Math Mammoth End-of-the-Year Test - Grade 5

The Four Operations

1. Solve (without a calculator).

   a. $1{,}035 \div 23$

   b. 492×832

2. Solve.

   a. $x - 56{,}409 = 240{,}021$

   b. $7{,}200 \div Y = 90$

   c. $N \div 14 = 236$

3. Write an equation to match this model, and solve it.

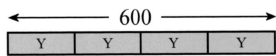

4. Place parentheses into the equations to make them true.

   a. $42 \times 10 = 10 - 4 \times 70$

   b. $143 = 13 \times 5 + 6$

5. Write a single expression (number sentence) for the problem, and solve.

> A store was selling movies that originally cost $19.95 with a $5 discount.
> Mia bought five of them. What was the total cost?

6. Is 991 divisible by 4?

   Why or why not?

7. Factor the following numbers to their prime factors.

a. 26 /\	b. 40 /\	c. 59 /\

Large Numbers

8. Write the numbers.

   a. 70 million 16 thousand 90

   b. 32 billion 232 thousand

9. Estimate the result of 31,933 × 305.

10. What is the value of the digit 8 in the number **56,782,010,000**?

11. Round these numbers to the nearest thousand, nearest ten thousand, nearest hundred thousand, and nearest million.

number	593,204	19,054,947
to the nearest 1,000		
to the nearest 10,000		
to the nearest 100,000		
to the nearest million		

Problem Solving

12. Jack has an 8-ft long board. He cuts off 1/6 of it.
    How long is the remaining piece, in feet and inches?

13. A website charges a fixed amount for each song download.
    If you can download six songs for $4.68, then how much would
    it cost to download ten songs?

14. A meal in a fancy restaurant costs three times as much as a meal in the cafeteria.
    The meal in the fancy restaurant costs $36. In a 5-day workweek, Mary ate lunch
    at the fancy restaurant once, and in the cafeteria the rest of the days.
    How much did she spend on lunch that week?

15. A blue swimsuit costs $42 and a red swimsuit costs 5/6 as much. How much would the two swimsuits cost together?

    Mark the $42 in the bar model. Mark what is not known with "?". Solve.

16. A bag has green and purple marbles. Two-fifths of the marbles are green, and the rest are purple.

    a. Draw a bar model for this situation.

    b. If there are 134 green marbles, how many are purple?

17. Karen and Ann share the cost of a DVD that costs $29.90 so that Karen pays 3/5 of it and Ann pays 2/5 of it.

    a. *Estimate* how much each person will pay.

    b. Find the exact amount of how much each person will pay.

Decimals

18. Write the decimals indicated by the arrows.

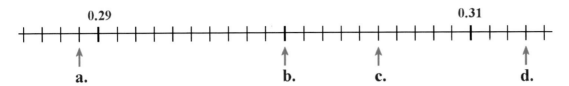

a. _____ b. _____ c. _____ d. _____

19. Complete.

a. $0.9 + 0.05 =$ _____	b. $0.28 +$ _____ $= 1$	c. $0.82 - 0.2 =$ _____
d. $1.3 - 0.04 =$ _____	e. $0.25 + 0.8 =$ _____	f. _____ $- 0.2 = 0.17$

20. Write as decimals.

a. $\dfrac{8}{100} =$

b. $\dfrac{81}{1000} =$

c. $5\dfrac{21}{100} =$

21. Write as fractions or mixed numbers.

a. 0.048   b. 1.004   c. 7.22

22. Compare, and write $<$ or $>$.

a. 0.31 ☐ 0.031   b. 0.43 ☐ 0.093   c. 1.6 ☐ 1.29

23. Round the numbers to the nearest one, nearest tenth, and nearest hundredth.

rounded to...	nearest one	nearest tenth	nearest hundredth	rounded to...	nearest one	nearest tenth	nearest hundredth
5.098				0.306			

24. Solve.

a. $0.4 \times 7 =$	d. $10 \times 0.05 =$	g. $1.1 \times 0.3 =$
b. $0.4 \times 0.7 =$	e. $100 \times 0.05 =$	h. $70 \times 0.9 =$
c. $0.4 \times 700 =$	f. $1000 \times 0.5 =$	i. $20 \times 0.09 =$

25. Divide.

a. $0.36 \div 6 =$	c. $3 \div 100 =$	e. $16 \div 10 =$
b. $5.6 \div 7 =$	d. $0.7 \div 10 =$	f. $71 \div 100 =$

26. Convert.

a. 0.2 m = _____ cm	b. 0.4 L = _____ ml	c. 56 oz = _____ lb _____ oz
37 cm = _____ m	3.5 kg = _____ g	74 in = _____ ft _____ in
2.9 km = _____ m	240 g = _____ kg	15 C = _____ qt _____ C

27. Two liters of ice cream is divided equally into nine bowls. Calculate, to the nearest milliliter, how much ice cream is in *two* bowls.

28. Calculate.

   a. $4.2 - 2.78$

   b. $71.40 \div 5$

   c. 2.2×6.4

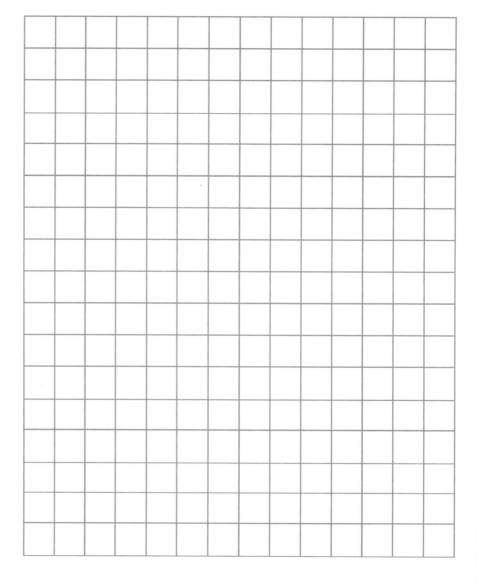

Graphs

29. Plot the points from the "number rule" on the coordinate grid.

    The rule for *x*-values:
    Start at 0, and add 1 each time.

    The rule for *y*-values:
    Start at 1, and add 2 each time.

x	0	1				
y	1					

30. In the grid draw a circle with a center point at (8, 4) and a radius of 3 units.

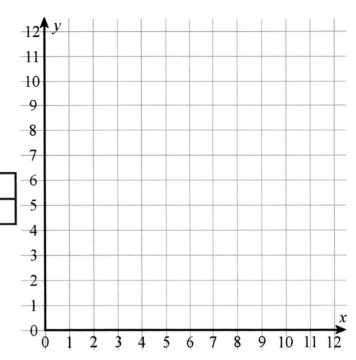

31. The table below gives the amount of sales in a grocery store from Monday through Friday.

Day	Sales (thousands of dollars)
Mon	125
Tue	114
Wed	118
Thu	130
Fri	158

    a. Make a line graph.

    b. Calculate the average daily sales for this period.

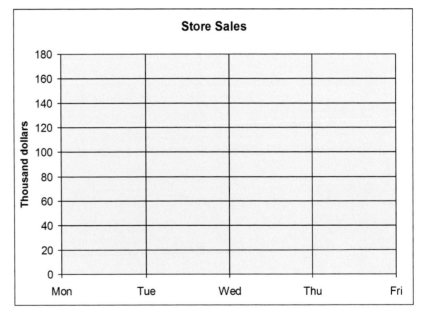

Fractions

32. Add and subtract.

a.	b.	c.
$3\frac{7}{9}$ $+\ 2\frac{5}{9}$ ——————	$5\frac{1}{6}$ $-\ 2\frac{5}{6}$ ——————	$3\frac{7}{10}$ $2\frac{8}{10}$ $+\ 7\frac{3}{10}$ ——————

33. Mark the fractions on the number line. $\frac{3}{4}$, $\frac{1}{3}$, $\frac{4}{6}$, $\frac{5}{12}$

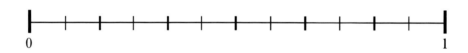

34. If you can find an equivalent fraction, write it. If you cannot, cross the whole problem out.

a. $\frac{5}{6} = \frac{}{20}$	b. $\frac{2}{7} = \frac{}{28}$	c. $\frac{3}{8} = \frac{15}{}$	d. $\frac{2}{9} = \frac{6}{}$

35. Find the errors in Mia's calculation and correct them.

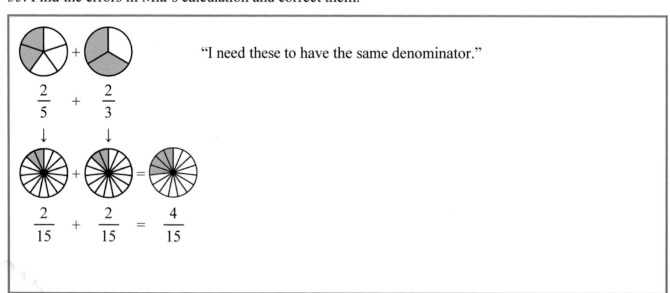

"I need these to have the same denominator."

36. Add and subtract the fractions and mixed numbers.

a. $\dfrac{1}{3} + \dfrac{5}{6}$	b. $\dfrac{4}{5} - \dfrac{1}{3}$
c. $6\dfrac{1}{8} - \dfrac{1}{2}$	d. $6\dfrac{7}{9} + 3\dfrac{1}{2}$

37. You need 2 3/4 cups of flour for one batch of rolls.
    Find how much flour you would need for three batches of rolls.

38. Compare the fractions, and write <, >, or = in the box.

   a. $\dfrac{6}{9} \square \dfrac{6}{13}$ b. $\dfrac{6}{13} \square \dfrac{1}{2}$ c. $\dfrac{5}{10} \square \dfrac{48}{100}$ d. $\dfrac{1}{4} \square \dfrac{25}{100}$ e. $\dfrac{5}{7} \square \dfrac{7}{10}$

39. Simplify the following fractions if possible. Give your answer as a mixed number when you can.

a. $\dfrac{21}{15} =$	b. $\dfrac{29}{36} =$	c. $\dfrac{42}{48} =$

40. Is the following multiplication correct? $\dfrac{2}{3} \times$ ⊘ $=$ ⊘
    If not, correct it.

41. Multiply the fractions, and shade a picture to illustrate the multiplication.

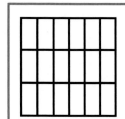 a. $\dfrac{1}{3} \times \dfrac{5}{6}$

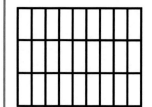 b. $\dfrac{2}{9} \times \dfrac{2}{3}$

42. How many 1/4 ft pieces can you cut from a string that is 15 feet long?

43. Three people share half a pizza evenly. What fractional part of the original pizza does each one get?

44. Solve. Give your answer as a mixed number and simplified to lowest terms.

a. $\dfrac{7}{6} \times 9$

b. $\dfrac{1}{7} \div 3$

c. $\dfrac{4}{5} \times 3\dfrac{2}{3}$

d. $2 \div \dfrac{1}{9}$

Geometry

45. Measure the sides of the triangle in inches. Find its perimeter.

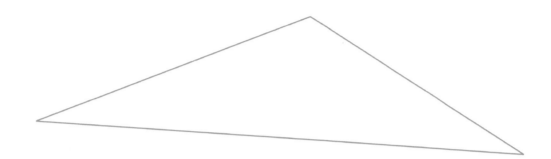

46. Below you see two triangles and two quadrilaterals. Classify the triangles according to their sides and angles. Name the quadrilaterals.

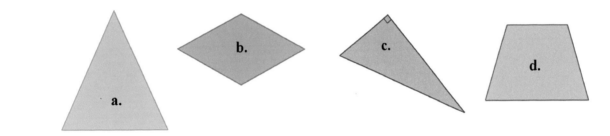

a. _____

b. _____

c. _____

d. _____

47. a. A square has a perimeter of 12 m. What is its area?

b. A square has an area of 25 ft^2. What is its perimeter?

48. Is a square a trapezoid? Why or why not?

49. Can an obtuse triangle be isosceles?
    If not, explain why not.
    If yes, sketch an example.

50. **a.** Draw a right triangle with 5 cm and 7 cm perpendicular sides.

    b. Find its perimeter.

    c. Measure its angles. They measure _____°, _____°, and _____°.

51. This is a rectangular prism.
    Find its volume.

    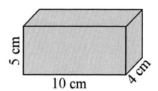

52. Matthew has a rainwater collection tank in his yard that is rectangular, like a box. It is 1.2 m long, 60 cm wide, and 1 m tall.

    a. Find the volume of the tank in cubic <u>meters</u>.

    b. After a rainy night, the tank was about 1/3 full.
    About how many liters of water were in the tank?
    1 cubic meter equals 1,000 liters.

Math Mammoth Grade 5 Review Workbook Answers

The Four Operations Review, p. 7

1. a. 281   b. 69   c. 95,118

2. 83,493 − 21,390 = 62,103

3. a. 55   b. 140   c. 30   d. 56

4. a. 606   b. 902   c. 810   d. 93   e. 1,201

5. a. 9   b. 3   c. 8

6. a. $x − 9$ b. $y + 3 + 8 = 28$ c. $60 ÷ b = 12$ d. $8 × x × y$

7. **(4)** 4 × $3.75 ÷ 3 = $5. Each girl paid $5.

8. a. (12 + 17) ÷ 2 = $14.50. Each paid $14.50.
   b. 5 × 4.50 − 2 = $20.50. Henry paid $20.50.

9. a. R ÷ 4 = 544; R = 2,176   b. 4 × R = 300; R = 75

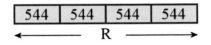   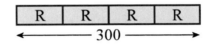

10.

Divisible by	2	3	5	6	9
534	X	X		X	
123		X			

Divisible by	2	3	5	6	9
1,605		X	X		
2,999					

11. a. 21 = 3 × 7   b. 12 = 2 × 2 × 3   c. 38 = 2 × 19
    d. 75 = 3 × 5 × 5   e. 124 = 2 × 2 × 31   f. 89 = 1 × 89

The Four Operations Test, p. 11

1. a. 56   b. 605   c. 185,725

2. Y = 28,451 (Add 8,687 and 19,764 to solve for *y*.)

3. a. 144   b. 76

4. a. 901   b. 311   b. 809

5. a. 31   b. 80

6. a. 11*s* or 11 × *s* or *s* × 11   b. 48/*b* = 8 or 48 ÷ *b* = 8

7. $20 − (5 × $2.50) = $7.50. Her change was $7.50.

8. a.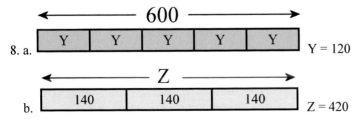
   Y = 120

   b. Z = 420

9. No, it is not. Since 990 is divisible by 3, 991 cannot be divide evenly. Or, when you add the digits of 991, you get 19, which is not divisible by 3, so 991 is not either.

10. a. 16 = 2 × 2 × 2 × 2   b. 34 = 2 × 17   c. 80 = 2 × 2 × 2 × 2 × 5

Large Numbers and the Calculator Review, p. 13

1. a. 9,070,560   b. 60,007,540   c. 50,000,050,050   d. 431,098,000,940

2.

a. 405,2<u>2</u>9,020 Place: ten thousands Value: twenty thousand	b. 97,02<u>4</u>,003,245 Place: one millions Value: four million
c. 2<u>3</u>0,560,079,000 Place: ten billions Value: thirty billion	d. 4,<u>5</u>89,211,000 Place: hundred million Value: five hundred million

3.

number	69,066	14,506,439	389,970,453	12,976,895,322
to the nearest 1,000	69,000	14,506,000	389,970,000	12,976,895,000
to the nearest 10,000	70,000	14,510,000	389,970,000	12,976,900,000
to the nearest 100,000	100,000	14,500,000	390,000,000	12,976,900,000
to the nearest million	0	15,000,000	390,000,000	12,977,000,000

4.

a. $8^2 = 8 \times 8 = 64$ b. $4^3 = 4 \times 4 \times 4 = 64$ c. $10^3 = 10 \times 10 \times 10 = 1,000$	d. $1^5 = 1 \times 1 \times 1 \times 1 \times 1 = 1$ e. $100^2 = 100 \times 100 = 10,000$ f. $2^5 = 2 \times 2 \times 2 \times 2 \times 2 = 32$

5.

a. $3 \times 3 \times 3 = 3^3 = 27$ b. $7 \times 7 = 7^2 = 49$ c. five squared $= 5^2 = 25$ d. ten cubed $= 10^3 = 1,000$	e. $10 \times 10 \times 10 \times 10 \times 10 = 10^5 = 100,000$ f. $2 \times 2 \times 2 \times 2 \times 2 \times 2 = 2^6 = 64$ g. five cubed $= 5^3 = 125$ h. ten to the sixth power $= 10^6 = 1,000,000$

6. a. 22,000,000   b. 3,600,000   c. 10,000,000,000   d. 6,000
   e. 800,000   f. 200,000,000   g. 210,000   h. 829,000,000

7. 8,079,083

8.

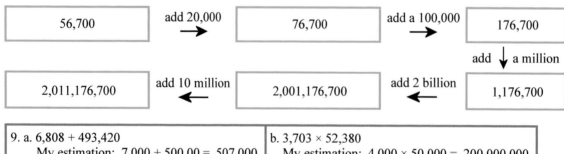

9.

a. 6,808 + 493,420 My estimation: <u>7,000 + 500,00 = 507,000</u> Exact answer: <u>500,228</u> Error of estimation: <u>6,772</u>	b. 3,703 × 52,380 My estimation: <u>4,000 × 50,000 = 200,000,000</u> Exact answer: <u>193,963,140</u> Error of estimation: <u>6,036,860</u>

10. $8,881,833,600 weekly

11. $36 billion or $36,000,000,000 (Multiply $9,000 and 4,000,000.)

Mystery Number: a. 2,260,430   b. 3,023,183

Large Numbers and the Calculator Test, p. 17

1. a. 70,006,324 b. 4,000,032,000 c. 98,089,000,098
2. a. 90,000 b. 30 million or 30,000,000 c. 4 billion or 4,000,000,000
3.

number	183,602	355,079,933	29,928,900
to the nearest 1,000	184,000	355,080,000	29,929,000
to the nearest 10,000	180,000	355,080,000	29,930,000
to the nearest 100,000	200,000	355,100,000	29,900,000
to the nearest million	0	355,000,000	30,000,000

4. a. 81 b. 1,000 c. 27
5. a. $6^2 = 36$ b. $2^5 = 32$
6. a. 36,000,000 b. 60,000,000 c. 70,000 d. 48,000,000

7.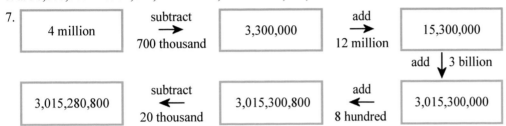

8. Estimations may vary.

a. 209,800 − 4,730	b. 2,543 × 5,187
Estimation: 210,000 − 5,000 = 205,000	Estimation: 2,500 × 5,000 = 12,500,000
Exact answer: 205,070	Also acceptable 3,000 × 5,000 = 15,000,000
Error of estimation: 70	Exact answer: 13,190,541
	Error of estimation: 690,541 or 1,809,459

c. 56,493,836 + 345,399 + 7,089,400

Estimation: 56,000,000 + 345,000 + 7,000,000 = 63,345,000
Also acceptable: 56,000,000 + 0 + 7,000,000 = 63,000,000
Exact answer: 63,928,635   Error of estimation: 583,635 or 928,635

9. A family of four would have to pay $9,200 for it. ($700,000,000,000 / 305,000,000 = $2,295.08)
   Hint: If the numbers do not fit onto your calculator's screen, remove the same amount of zeros from both the dividend and the divisor. Then divide. In other words, $700,000,000,000 ÷ 305,000,000 becomes $700,000 ÷ 305, and the two problems have the same answer.

Mixed Review 1, p. 19

1.

a. 76 − 65 = 11 subtrahend	b. 57 − 39 = 18 minuend	c. 48 − 29 = 19 difference

2. $x + 9,380 + 3,928 = 93,450$; $x = 80,142$
3. a. 84,000 b. 132,000,000 c. 300,000,000 d. 10,000 e. 27 f. 7,000,000
4. 10^6
5. There are 8,760 hours in a year.
   Estimate: 360 × 30 = 10,800
6. a. 128; 43 × 128 = 5,504   b. 95; 82 × 95 = 7,790
7. a. 2 × 2 × 7 b. 2 × 7 × 7 c. 2 × 3 × 11 d. 1 × 17 e. 3 × 17 f. 1 × 53

Mixed Review 2, p. 21

1. a. $90 + (70 + 80) \times 2 = 390$ b. $378 = 6 \times (8 + 13) \times 3$ c. $90 \times 4 = (180 - 60) \times 3$

2. a. $x = 20$

   | x | x | x | x | 120 |

   (total 200)

   b. $x = 9$

   | 25 | x | x | x |

   (total 52)

3. a. 630   b. 322

4. a. He earned $480 in one week.   b. They raced 1,496 miles.

5. **(3)** $50 - ($9 + $9 + $9 + $9)$ and **(5)** $50 - 4 \times $9 = 14. His change was $14.

6. Estimates may vary. The method I use most often in estimating multiplications is to round one number up, the other down, to numbers that are easy to multiply mentally. For example, in 173×35, rounding to 200×30 may look "unorthodox", but it gives a good estimate.

a.
```
    173
 ×   35
    865
   5190
   6055
```
Estimate: 200×30
$= 6,000$

Error of estimation: 55

b.
```
    269
 ×  537
   1883
   8070
 134500
 144453
```
Estimate: 300×500
$= 150,000$

Error of estimation: 5,547

c.
```
    892
 ×  340
      0
  35680
 267600
 303280
```
Estimate: 900×300
$= 270,000$

Error of estimation: 33,280

Problem Solving Review, p. 23

1. a. 211   b. 311

2. a. $x = (164 - 72) \div 2 = 46$

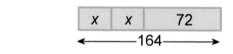

  b. $x = (1080 - 420) \div 5 = 132$

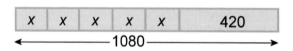

3. $109 - $25 - $10 = 74

| 10 | 25 | x |
← 109 →

Problem Solving Review, cont.

4. a. (137 qt − 45 qt) ÷ 2 = 46 qt. Eva canned 46 quarts.

   b. Joe's raft was 3 ft longer than Jay's, which is 1/3 of Joe's raft.
      One block in the model is therefore 3 ft. Jay's raft was 6 feet long.

   c. (112 mi + 35 mi) × 2 = 294 mi

   d. One light bulb costs $7.50 ÷ 5 = $1.50. Eight would cost 8 × $1.50 = $12.00.

5. ($175 − $37) ÷ 2 = $69. Austin was given $69.

6. The discounted price is 3/4 of the normal price. Three-fourths of $364 is $364 ÷ 4 × 3 = $273.

7. One block in the model is 69 ÷ 3 = 23. The chess club has 4 × 23 = 92 members.

Problem Solving Test, p. 27

1.

a. Equation: $4x = 32 + 4$ ($4x = 36$) Solution: $x = 9$	b. Equation: $x + 47 = 2x$ ($47 = x$) Solution: $x = 47$

2. a. $2x + 100 = 302$. Solution: $x = 101$. (Subtract 100 from 302, then divide the result by 2.)

   b. $4x + 442 = 998$. Solution: $x = 139$. (Subtract 442 from 998, then divide the result by 4.)

3. The phones cost $120. (The discount is 1/6 of $48, or $8. So, one discounted phone costs $40, and three cost $120.)

4. The younger sister gets 109 rocks.
   First subtract 250 − 32 = 218 to get the amount they both would have if the "extra" 32 weren't there. Then divide that by two to get 109.

5. a. Two kilograms cost $3. (One kilogram costs $7.50 ÷ 5 = $1.50.)
   b. Henry's change is $7.

6. Since the high-quality drive costs 3 times as much as the low-quality one, the bar model has *four* parts.
   One part is $820 ÷ 4 = $205. The low-quality hard drive costs $205.

7. One block in the bar model is 66 cm ÷ 3 = 22 cm.
   Dad is therefore 8 × 22 cm = 176 cm tall.

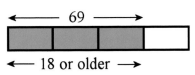
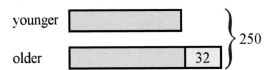
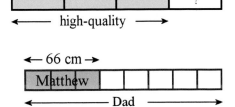

Mixed Review 3, p. 29

1. $x = 547 - 119 - 38 = 390$. See the bar model on the right.

2. 3×14 gal $\times 52 = 2{,}184$ gallons

3. a. $16 - 7 = 9$ b. $3 + 9 + y = 20$

4. a. 14   b. 10   c. 71

5. c. $7 \times 65 + 3$

6. a. the same   b. not the same   c. not the same

7. a. $64 = 2 \times 2 \times 2 \times 2 \times 2 \times 2$
   b. $60 = 2 \times 2 \times 3 \times 5$
   c. $85 = 5 \times 17$

8.

$2 \times 79 = 158$
$3 \times 79 = 237$
$4 \times 79 = 316$
$5 \times 79 = 395$
$6 \times 79 = 474$
$7 \times 79 = 553$
$8 \times 79 = 632$
$9 \times 79 = 711$
	`    1 1 3`
	`79) 8 9 2 7`
	`    7 9`
	`    1 0 2`
	`    - 7 9`
	`      2 3 7`
	`    - 2 3 7`
	`          0`
		Check:
		`  1 1 3`
		`×   7 9`
		`──────`
		`1 0 1 7`
		`7 9 1 0`
		`──────`
		`8 9 2 7`

9.

a. $2 \times 10^4 = 20{,}000$	b. $712 \times 10^3 = 712{,}000$	c. $55 \times 10^6 = 55{,}000{,}000$
d. $6 \times 10^3 = 6{,}000$	e. $18 \times 10^7 = 180{,}000{,}000$	f. $69 \times 10^6 = 69{,}000{,}000$

10.

a. $15{,}278 \times 3{,}892$ (round to thousands) My estimation: $15{,}000 \times 4{,}000 = 60{,}000{,}000$ Exact answer: $59{,}461{,}976$ Error of estimation: $538{,}024$	b. $19{,}945{,}020 - 6{,}320{,}653$ (round to millions) My estimation: $20{,}000{,}000 - 6{,}000{,}000 = 14{,}000{,}000$ Exact answer: $13{,}624{,}367$ Error of estimation: $375{,}633$

Mixed Review 4, p. 31

1. 389

2. a. 124   b. 84   c. 55   d. 490

3. a. 26, 28   b. 47, 135   c. 450, 600

4. a. 78,000,016,038   b. 844,012,000,704

5.

number	32,274,302	64,321,973	388,491,562	2,506,811,739
to the nearest 1,000	32,274,000	64,322,000	388,492,000	2,506,812,000
to the nearest 10,000	32,270,000	64,320,000	388,490,000	2,506,810,000
to the nearest 100,000	32,300,000	64,300,000	388,500,000	2,506,800,000
to the nearest million	32,000,000	64,000,000	388,000,000	2,507,000,000

Mixed Review 4, cont.

6.

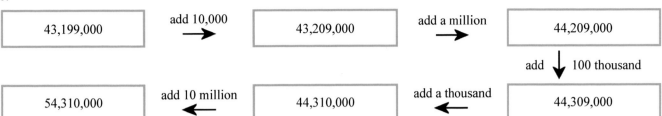

7. a. 5 × 6 + 50   b. 10 − (9 − 6)

8. (21 × 2) + (20 × 1.50) + 12 = $84 total cost.

9. a. 4,815,598,182   b. 1,046,556,957

10. Estimates may vary.

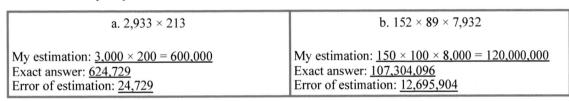

Decimals Review, p. 33

1.

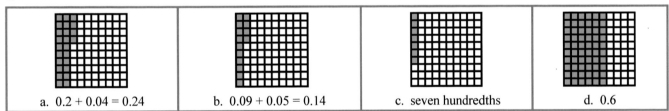

2. a. 0.495 = 4 × (1/10) + 9 × (1/100) + 5 × (1/1000)

   b. 2.67 = 2 × 1 + 6 × (1/10) + 7 × (1/100)

3. a. 0.042   b. 0.047   c. 0.05   d. 0.055   e. 0.062

4. a. >   b. >   c. <   d. >   e. <

5. a. 0.03   b. 0.048   c. 1.209   d. 3.39

6. a. $1\frac{3}{10}$ b. $2\frac{15}{100}$ c. $\frac{8}{1000}$ d. $\frac{38}{1000}$

7.

rounded to...	nearest one	nearest tenth	nearest hundredth
4.608	5	4.6	4.61
3.109	3	3.1	3.11
2.299	2	2.3	2.30
0.048	0	0.0	0.05

Decimals Review, cont.

8.

a. $0.3 + 0.005 = 0.305$ $0.03 + 0.5 = 0.53$	b. $0.9 - 0.7 = 0.2$ $0.9 - 0.07 = 0.83$
c. $0.008 + 0.9 + 5 = 5.908$ $0.9 + 0.8 + 0.17 = 1.87$	d. $2.5 - 1.02 = 1.48$ $7.8 - 0.9 - 0.04 = 6.86$

9. a. $0.21 + 0.79 = 1$ b. $0.004 + 0.996 = 1$ c. $4.391 + 0.609 = 5$

10. a. 3.944   b. 0.099

11. a. Each child should have about $3.00 left.   b. $25 ÷ 5 - $2.05   c. Each child has $2.95 left.

12.

a. $0.4 \times 8 = 3.2$ b. $6 \times 0.009 = 0.054$	c. $20 \times 0.5 = 10$ d. $100 \times 0.3 = 30$	e. $0.9 \times 0.2 = 0.18$ f. $0.06 \times 0.3 = 0.018$

13.

a. $0.35 \div 5 = 0.07$ b. $4.5 \div 9 = 0.5$	c. $0.4 \div 10 = 0.04$ d. $5 \div 100 = 0.05$	e. $0.38 \div 10 = 0.038$ f. $7 \div 1000 = 0.007$

14.

a. $0.8 \times 0.5 = 0.40$ b. $8 \times 0.008 = 0.064$	c. $7 \times 0.5 = 3.5$ d. $0.6 \times 0.04 = 0.024$	e. $0.9 \times 8 = 7.2$ f. $9 \times 0.09 = 0.81$

15.

a. $0.07 \times 10^2 = 7$ $10^5 \times 1.08 = 108,000$	b. $3,300 \div 10^4 = 0.33$ $239.8 \div 10^3 = 0.2398$

16. a. 0.7×5 kg $= 3.5$ kg   b. 0.06×1.2 m $= 0.072$ m   c. 0.35×2 L $= 0.7$ L

17. a. 1.817   b. 0.355   c. 0.85

18. Answers will vary. See some example answers below:
    The answer is less than 0.7 because you multiply 0.7 by a number that is less than one.
    The answer has to be less than 0.7 because multiplying by 0.4 means you are taking a fractional part of 0.7.
    Multiplying by 0.4 means taking 4/10 part of 0.7, and 4/10 is less than 1, so the answer is less than 0.7

19. The discounted price is $100.80. For example, you can calculate $0.8 \times \$126 = \$100.80$.

20. a. 382. The division $152.8 \div 0.4$ is converted into $1,528 \div 4 = 382$.
    b. 34.7. The division $2.776 \div 0.08$ is converted into $27.76 \div 0.8$ and then into $277.6 \div 8 = 34.7$.
    c. 163.64. The division $180 \div 1.1$ is converted into $1800 \div 11 \approx 163.636$.
    d. 0.29. The division $2.000 \div 7 = 0.285$ R0.005.

21.

a. 0.9 m = 90 cm 45 cm = 0.45 m 1.5 km = 1,500 m	b. 0.6 L = 600 ml 5,694 ml = 5.694 L 0.09 L = 90 ml	c. 2.2 kg = 2,200 g 390 g = 0.390 kg 0.02 kg = 20 g

22.

a. 6 ft 11 in. = 83 in. 3 lb 11 oz = 59 oz 3 C = 24 oz	b. 2 gal = 32 C 5 qt = 10 pt 54 oz = 6 C 6 oz	c. 78 oz = 4 lb 14 oz 39 in = 3 ft 3 in 102 in = 8 ft 6 in

Decimals Review, cont.

23.

a. 2.65 mi = 13,992 ft 10.9 mi = 19,184 yd	b. 3,800 ft = 0.72 mi 3,500 yd = 1.99 mi	c. 4.54 lb = 72.64 oz 10.2 ft = 122.4 in

24. a. Each box weighs 5.2 kg or 5 1/5 kg. 26.0 kg ÷ 5 = 5.2 kg or 26 kg ÷ 5 = 5 1/5 kg
    b. They cost $15.60 per box. 5.2 × 3 = 15.6.

25. Edward earns $446.50. Multiply 38 × $11.75 = $446.50 to get that. He pays in taxes $446.50 ÷ 5 = $89.30. He takes home $357.20 after taxes.

26. The smaller pitcher holds 1.55 liters and the larger holds 2.1 liters. (3.65 L − 0.55 L) ÷ 2 = 3.1 L ÷ 2 = 1.55 L.

Decimals Test, p. 39

1. a. 0.088
   b. 0.091
   c. 0.10 or 0.1
   d. 0.107
   e. 0.112

2. a. 5.7   b. 0.24   c. 0.35   d. 0.02

3. a. 0.21   b. 0.046   c. 3.07   d. 20.2

4. a. $\frac{6}{10}$ b. $\frac{82}{100}$ c. $1\frac{208}{1000}$ d. $\frac{93}{1000}$

5. a. <   b. >   c. <   d. <

6.

rounded to...	nearest one	nearest tenth	nearest hundredth	rounded to...	nearest one	nearest tenth	nearest hundredth
8.816	9	8.8	8.82	0.398	0	0.4	0.40
1.495	1	1.5	1.50	9.035	9	9.0	9.04

7. a. 2.8   b. 0.63   c. 1   d. 9   e. 200   f. 0.64

8. a. 0.04   b. 0.009   c. 0.02   d. 0.08   e. 0.043   f. 0.007

9. a. 500   b. 780,000   c. 0.035   d. 1.32

10. 1.211 (The sum is 1.109 + 0.102 = 1.211)

11. a. 0.6 × 20 = 12   b. 13.08

12. a. 1.306   b. 5.25

13. Answers will vary. See some example answers below:
    The answer is less than 0.8 because you multiply 0.8 by a number that is less than one.
    The answer has to be less than 0.8 because multiplying by 0.9 means you are taking a fractional part (9/10) of 0.8.
    Multiplying by 0.9 means taking 9/10 part of 0.8, and 9/10 is less than 1, so the answer is less than 0.8.
    The answer is also less than 0.9.

14. Each box weighs 7 kg ÷ 4 = 7/4 kg = 1 3/4 kg (*or* 1.75 kg *or* 1 kg 750 g *or* 1,750 g).

15.

a. 0.7 m = 70 cm 3.2 km = 3,200 m	b. 2,650 ml = 2.65 L 0.9 L = 900 ml	c. 5.16 kg = 5,160 g 400 g = 0.4 kg

Decimal Test, cont.

16.

a. 8 ft 10 in. = 106 in. 183 in. = 15 ft 3 in	b. 2 gal 3 C = 35 C 45 oz = 5 C 5 oz	c. 81 oz = 5 lb 1 oz 165 oz = 10 lb 5 oz

17. 1.6 L *or* 1,600 ml. (0.9 L + 350 ml + 350 ml = 0.9 L + 0.7 L = 1.6 L)

18. a. $1.12. First, find the kilogram price. Since 2 kg costs $4.48, 1 kg costs $2.24, and 1/2 kg costs $1.12.

    b. She should charge $0.56. Simply find 1/2 of the price for 1/2 kg, which was $1.12.

19. Two discounted DVDs cost $23.94. First, find the price of one discounted CD: $19.95 ÷ 5 × 3 = $11.97. Then multiply that by 2.

20. a. b. 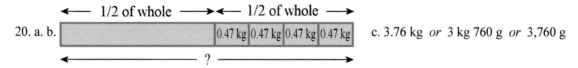 c. 3.76 kg *or* 3 kg 760 g *or* 3,760 g

Mixed Review 5, p. 43

1. a. 1,070   b. 2,515   c. 901

2. a. *y* = 62,103 (subtract 21,390 from 83,493)   b. *s* = 317 (divide 6,340 by 20)

3. a. 10   b. 30   c. 58   d. 146

4. a. 15,000,600,024   b. 42,080,017

5. a. 1, 2, 3, 6, 7, 14, 21, 42   b. 1, 2, 4, 8, 16, 32, 64

6. a. Twenty song downloads would cost $2.20. $5.50 ÷ 50 × 20 = $2.20.

    b. Mr. Doe had $585 left. $870 − ($870 ÷ 6) − $140 = $585

    c. Henry owns 4 × 450 = 1,800 stamps.

    d. The cheaper hammer cost ($64 − $28) ÷ 2 = $18.

Mystery number: 264

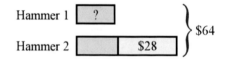

Mixed Review 6, p. 45

1. Together they earned $225. Jack's sister earns $125 ÷ 5 × 4 = $100.

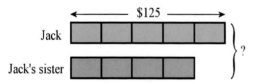

2. The price difference is $2.60.
   The cheaper thermometer costs $10.40 ÷ 4 × 3 = $7.80. Then the difference is $10.40 − $7.80 = $2.60.
   You can also solve this by thinking that the difference in prices is 1/4 of the price of the expensive thermometer, and $10.40 ÷ 4 = $2.60.

3. 120 ÷ 5 × 2 = 48; 120 − 48 = 72. She has 72 more marbles.

4. a. Eighteen students are new. Simply find 6/50 of 150: 1/50 of 150 is 3, so 6/50 is six times as much, or 18.
   b. Twelve students have studied English before. First find 1/3 of 18 students, which is 6.
      The rest, or 12, have studied English before.

5. 182

Mixed Review 6, cont.

6. a. M = 50   b. M = 48   c. M = 90   d. N = 8,000   e. N = 21,000   f. N = 80

7. a. 350 ÷ x = 5   b. (15 − 6) + 8 or 8 + (15 − 6) or the same expressions without parentheses.

8. a. 3   b. 11   c. 4

9.

number	97,302	25,096,199	709,383,121	89,534,890,066
to the nearest 1,000	97,000	25,096,000	709,383,000	89,534,890,000
to the nearest 10,000	100,000	25,100,000	709,380,000	89,534,890,000
to the nearest 100,000	100,000	25,100,000	709,400,000	89,534,900,000
to the nearest million	0	25,000,000	709,000,000	89,535,000,000

Graphing and Statistics Review, p. 47

1. The rule is: $y = 9 - x$.

x	0	1	2	3	4
y	9	8	7	6	5

x	5	6	7	8	9
y	4	3	2	1	0

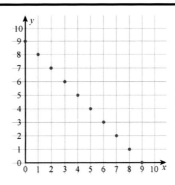

2. The mean is 11.67; the mode is 10.

3. a. Estimate: 3,750,000 tractors in 2010.
   b. From 1940 to 1950, the increase was about 1,750,000 tractors
   c. Slowly declining (but at a slightly increasing rate of decline).
   d. In 1930 there were about 1 million tractors; in 1960 about 4 1/2 million. So the increase was 4 1/2-fold.

4. a. In 2007: June, July, August, and November.
      In 2008: March, May, July, August, and November.
   b. June.

5. a.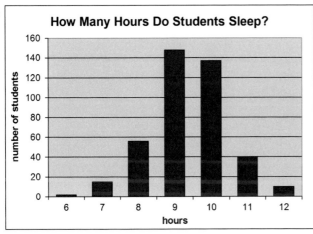

   b. The mode is 9 hours.
   c. The latter (6, 10, 8, 8, 9, 7, 11, 10, 9, 10, 11,...)
   d. 3827 hours ÷ 408 students = 9.38 hours/student ≈ 9.4 hours

Graphing and Statistics Test, p. 49

1. The coordinates of the other points are (5, 3).

2.

x	1	2	3	4
y	1	3	5	7

x	5	6	7	8
y	9	11	13	15

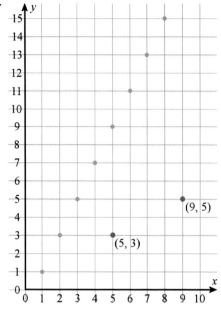

3. a. b.

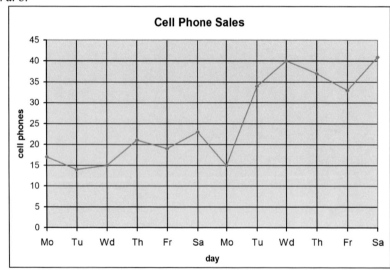

c. The promotion started on the second Tuesday.

Graphing and Statistics Test, cont.

4.
a.

Age	frequency
8-9	5
10-11	8
12-13	4
14-15	3

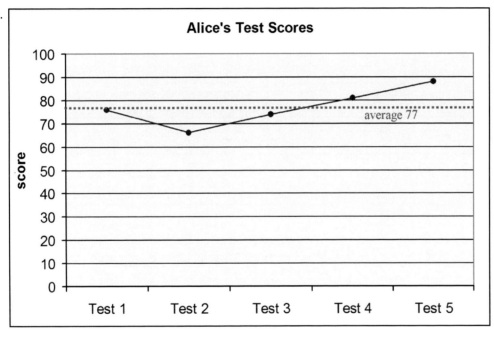

b. The mode is 10.

c. The average is 11.05.

5. a. & c.

<!-- Alice's Test Scores graph -->

b. 77   d. 79.75

Mixed Review 7, p. 51

1. a. 1.289   b. 3.108

2.

+ 0.05	+ 0.1	− 0.26	+ 0.01	+ 0.072	+ 0.028	− 0.19	− 0.01
2.2 2.25	2.35	2.09	2.1	2.172	2.2	2.01	2

3.

a.	b.	c.	d.
$2 \times 0.06 = 0.12$	$0.4 \times 0.7 = 0.28$	$100 \times 0.12 = 12$	$1.1 \times 0.9 = 0.99$
$2 \times 0.6 = 1.2$	$5 \times 0.007 = 0.035$	$0.5 \times 0.03 = 0.015$	$1000 \times 0.05 = 50$

Mixed Review 7, cont.

4. a. Estimate: 4 + 3 + 11 + 2 + 8 ≈ $28   b. Exact total: $28.50   c. Error of estimation: $0.50

5. a. 48 = 2 × 2 × 2 × 2 × 3   b. 71 is a prime number   c. 93 = 3 × 31

6. x = 5.62

7. a. 20,000   b. 9,000,000   c. 17,000

8. x ÷ 52 = 210; x = 10,920

9. (46 − 6) ÷ 2 = $20; Luisa spent $20 and Mary spent $26.

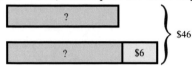

10. Half of John's money is: 2 × $48 + $120 = $216. Therefore, he earned 2 × $216 = $432.

11. Estimations may vary. For example: 12 × 235 cm ≈ 10 × 240 cm = 2,400 cm = 24 m.

12. a. 104 Check: 104 × 38 = 3,952   b. 15.8 Check: 15.8 × 17 = 268.6

13. a. 6.22   b. The division 5.175 ÷ 0.5 becomes 51.75 ÷ 5 = 10.35.

14.

a. 127,285 + 84,662 (round to thousands) My estimation: 127,000 + 85,000 = 212,000 Exact answer: 211,947 Error of estimation: 53	b. 12,705,143 − 6,460,788 (round to millions) My estimation: 13,000,000 − 6,000,000 = 7,000,000 Exact answer: 6,244,355 Error of estimation: 755,645

Mixed Review 8, p. 54

1.

Round this to the nearest →	unit (one)	tenth	hundredth	Round this to the nearest →	unit (one)	tenth	hundredth
4.925	5	4.9	4.93	5.992	6	6.0	5.99
6.469	6	6.5	6.47	9.809	10	9.8	9.81

2. 56 × 2 − 14 = 98

3. a. 0.37   b. 0.192   c. 0.328   d. 1.45   e. 1.05   f. 0.506

4. a. 3.09   b. 8.075

5.

a. 0.5 m = 50 cm 0.06 m = 6 cm 2.2 km = 2,200 m	b. 4.2 L = 4,200 mL 400 mL = 0.4 L 5,400 g = 5.4 kg	c. 800 g = 0.8 kg 4,550 m = 4.55 km 2.88 kg = 2880 g

6. One package of AAA batteries costs $1.90. To solve this, first subtract the cost of the AA batteries from the total: $17.04 − $5.64 = $11.40. Then divide that by six: $11.40 ÷ 6 = $1.90.

7. a. 5 R1, 5.17   b. 10 R3, 10.75

Mixed Review 8, cont.

8. a. Estimate: 2 × 12 = 24.   Exact: 24.035
   b. Estimate: 70 × 2 = 140.   Exact: 156.22
   c. Estimate: 7 × 3 = 21.   Exact: 21.442

9. a. One packet of seeds cost $1.89   b. One plant cost $1.92.
   c. The total cost was $26.67.

10. a. 0.7   b. 63   c. 29   d. 0.08   e. 0.045   f. 0.076

Fractions: Add and Subtract Review, p. 57

1. a. 19/2   b. 61/11   c. 58/7   d. 506/100

2. a. 4 1/10   b. 6 1/3   c. 3 1/9   d. 2 8/12

3. 23/6 = 3 5/6

4. a. 5 5/8   b. 6 12/20   c. 5 4/15

5. a. 15/21 + 7/21 = 22/21 = 1 1/21   b. 9/30 + 10/30 = 19/30
   c. 2 9/7 − 1 6/7 = 1 3/7   d. 2 16/20 + 3 5/20 = 5 21/20 = 6 1/20

6. a. <   b. <   c. =   d. <   e. <   f. >   g. <   h. <

7. From the first piece, she has left: 5 1/2 ft − 3 1/8 ft = 5 4/8 ft − 3 1/8 ft = 2 3/8 ft.
   From the second piece, she has left: 4 1/4 ft − 3 1/8 ft = 4 2/8 ft − 3 1/8 ft = 1 1/8 ft.
   Combined, those two pieces are 2 3/8 ft + 1 1/8 ft = 3 4/8 ft = 3 1/2 ft.

8. 1 − 32/100 − 42/100 − 2/10 = 1 − 32/100 − 42/100 − 20/100 = 6/100. So, 6/100 of the land is resting.

9. One-fifth of $35 is $7, so the discounted price would be $28. And 2/11 of $33 is $6, so its discounted price would be $27. So, 2/11 off of the $33-book is the better deal.
   If both books cost $50, then 1/5 off of it would be the better buy. This is because 1/5 is more than 2/11.

Fractions: Add and Subtract Test, p. 59

1. a. 8 2/3   b. 6 3/7   c. 6 4/5

2. a. 10 3/8   b. 2 2/5   c. 16 8/11

3. a. 2 5/7   b. 2 7/9   c. 4 9/15

4.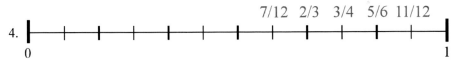

5.

| a. $\frac{3}{7} = \frac{9}{21}$ | b. $\frac{4}{3} = \frac{24}{18}$ | c. $\frac{5}{6} = \frac{}{11}$ NOT POSSIBLE | d. $\frac{2}{5} = \frac{8}{20}$ | e. $\frac{5}{6} = \frac{15}{18}$ |

6.

| a. $\frac{7}{4} > \frac{5}{3}$ | b. $\frac{5}{11} < \frac{1}{2}$ | c. $\frac{7}{10} > \frac{69}{100}$ | d. $\frac{3}{4} = \frac{75}{100}$ | e. $\frac{8}{7} > \frac{7}{9}$ |

7. We split 1/3 into additional pieces so it becomes 5/15, and similarly, we split 2/5 into additional pieces so they become 6/15. Now we can add. The answer is 11/15.

8. a. 1 5/12   b. 1/6   c. 2 9/14   d. 10 3/40

Fractions: Add and Subtract Test, cont.

9. $\frac{1}{2}, \frac{5}{9}, \frac{4}{7}, \frac{7}{5}$

10. 1/4 + 21/28 = 1 *or* 2/4 + 14/28 = 1 *or* 3/4 + 7/28 = 1

11. The sides measure 3 3/16 inches., 1 5/16 inches., and 2 1/4 inches. The perimeter is 6 3/4 inches.

Mixed Review 9, p. 61

1.

a. 452,9**1**2,980 Place: ten thousands place Value: 10,000 (ten thousand)	b. **6**,219,455,221 Place: billions place Value: 6,000,000,000 (six billion)

2. 60 × 60 = 3,600 seconds in an hour; 3,600 × 24 = 86,400 seconds in a day

3. (52 − 12) × 5 × 5 = 1,000. They do 1,000 hours of schoolwork in a year.

4. Estimate: 2 × 7 = 14. Exact answer: 14.348

5. a. 0.034; 0.021   b. 9.8; 46,700   c. 0.019; 300

6.

a. 5,070 g = _5.07_ kg 2.5 kg = _2,500_ g	b. 0.6 L = _600_ ml 10,500 ml = _10.5_ L	c. 0.06 km = _60_ m 2,600 m = _2.6_ km

7. a. 82.50 ÷ 0.06 = 825 ÷ 0.6 = 8,250 ÷ 6 = 1,375   b. 48.302 ÷ 0.2 = 483.02 ÷ 2 = 241.51

8. 3 × 2.40 − 0.15 + 0.30 = $7.35 for the three cups of yogurt.

9. $6.29 × 3 = $18.87 is the price of Shirt B.

10. Multiplying as if there was no decimal point, I get 1,000 × _7_. That equals _7,000_.
    Then, since my answer has to have thousandths, it needs _3_ decimal digits.
    So, the final answer is _7.000 or 7_.

11. a.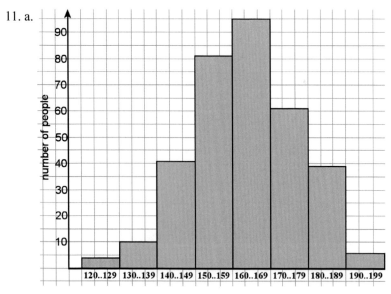

    b. 14 people   c. 45 people   d. About 82 + 41 + 10 + 4 = 137 children, about 95 + 61 + 39 + 6 = 201 adults
    e. It could be a group of people that were at the swimming pool at 5 pm on a certain Tuesday because there were both children and adults.

Mixed Review 10, p. 64

1. Estimates will vary.
    a. Estimate 300 × 280 = 84,000.   Exact 80,330.
    b. Estimate 530 × 400 = 212,000.   Exact 218,400.
    c. Estimate 900 × 200 = 180,000.   Exact 166,842.

2. 2,400 m − 2 × 250 m = 4,300 m or 4.3 km

3. 100 − 10 ÷ 2 = $45; Angi got $55 and Rebekkah got $45.

4. a. 0.4   b. 1.2   c. 30   d. 0.2   e. 20   f. 4

5. a. $y = 1.67$ b. $z = 1.681$

6. The mean is 18.

7. a. 0.908 = 9 × (1/10) + 0 × (1/100) + 8 × (1/1000)   b. 543.2 = 5 × 100 + 4 × 10 + 3 × 1 + 2 × (1/10)

8. a. 0.4   b. 0.06   c. 5   d. 0.04   e. 9   f. 20

9. Each student's share was $2,676.

10. A half of a gallon is eight cups, so there are six cups of milk left.

11. Sixty-one inches is five feet and one inch, so Eva is five inches taller than Ava.

12. 36 × 6 oz = 216 oz; 216 oz ÷ 16 oz = 13 1/2. The total weight is 13 lb 8 oz.

13. a. The histograms can vary based on how the bins are chosen. The example answer below uses a bin width of 10, starting at 0. One could also use a bin width of 11, starting at 4, or some other possibilities.

distance	frequency
0...9	5
10...19	6
20...29	3
30...39	1
40...49	1

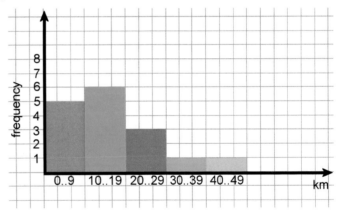

b. 17.1875 km   c. There are three modes: 7 km, 15 km, and 25 km.

Fractions: Multiply and Divide Review, p. 67

1.

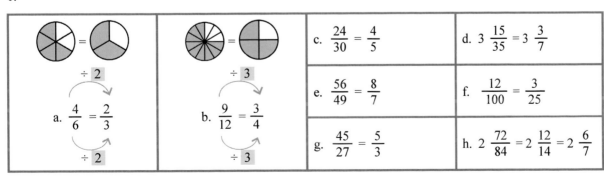

Fractions: Multiply and Divide, cont.

2.

a. $3 \times 1\frac{1}{3} = 3\frac{3}{3} = 4$

b. $2 \times \frac{5}{6} = \frac{10}{6} = 1\frac{4}{6} = 1\frac{2}{3}$

3. a. 2 4/5   b. 5/21   c. 17 1/5   d. 6 11/18

4.

a. $\frac{\cancel{7}^1}{\cancel{14}_2} \times \frac{\cancel{3}^1}{\cancel{12}_4} = \frac{1}{8}$

b. $\frac{\cancel{5}^1}{\cancel{24}_2} \times \frac{\cancel{12}^1}{\cancel{30}_6} = \frac{1}{12}$

5.

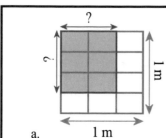

a. Side lengths: $\frac{2}{3}$ m and $\frac{3}{4}$ m

Area: $\frac{2}{3}$ m $\times \frac{3}{4}$ m $= \frac{6}{12}$ m² $= \frac{1}{2}$ m²

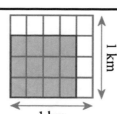

b. Side lengths: $\frac{4}{5}$ km and $\frac{3}{4}$ km

Area: $\frac{4}{5}$ km $\times \frac{3}{4}$ km $= \frac{12}{20}$ km² $= \frac{3}{5}$ km²

6.

a. $\frac{5}{6}$ m $\times \frac{1}{2}$ m $= \frac{5}{12}$ m²

b. $\frac{2}{3}$ in. $\times \frac{1}{6}$ in. $= \frac{2}{18}$ in² $= \frac{1}{9}$ in²

7. 5 × (3/4 mi) = 15/4 mi = 3 3/4 mi

8. a. Four pieces are left. Dog ate 2/3 of the 12 pieces, so 1/3 of the 12 pieces are left, which is 4 pieces.
   b. 4/20 = 1/5 of the pie is left now.

9.

a. $1 \div 3 = \frac{1}{3}$

b. $\frac{1}{2} \div 3 = \frac{1}{6}$

Fractions: Multiply and Divide Review, cont.

10. a. 6   b. 16   c. 1/10   d. 1/21   e. 2 1/4   f. 1/16   g. 2/3   h. 2/10   i. 1/4

11. a. Each piece is 1 3/4 in. long: 7 in. ÷ 4 = 7/4 in. = 1 3/4 in.
    b. (4/5) × (1/3) = 4/15
    c. 11 lb ÷ 5 = 11/5 lb = 2 1/5 lb
    d. (3/4) × $24 = $18
    e. Seventy were not women. First, find 1/8 of 112, which is 14.
       Then, since 5/8 were not women, multiply that by 5 to get 70.

12. a. <   b. >   c. =

13. 1 − 1/3 − 1/4 = 1 − 4/12 − 3/12 = 5/12. So, 5/12 of the cake was decorated with strawberry frosting.

14. After a day, 5/6 of 30 slices were left, which is 25 slices. The family ate 1/5 of 25 slices, which is 5 slices.
    Therefore, 20 slices are now left.

15. Note: 9/16 cup or 9/16 teaspoon is not commonly found on measuring cups. You would just use a tad over 1/2 C or 1/2 teaspoon.

    Brownies

    2 1/4 cups sweetened carob chips
    6 tablespoons olive oil
    2 small eggs
    3/8 cup honey
    3/4 teaspoon vanilla
    9/16 cup whole wheat flour
    9/16 teaspoon baking powder
    3/4 cup walnuts or other nuts

16. a. The 6 ½ in. by 8½ in. sheet has a greater area.
       The area of the 6 ½ in. by 8½ in. sheet is (6 1/2 in.) × (8 1/2 in.) = (13/2 in.) × (17/2 in.) = 221/4 in² = 55 ¼ in².
       The area of the 5¾ in. by 9 in. sheet. is (5 3/4 in.) × 9 in = (23/4 in.) × 9 = 207/4 in² = 51 ¾ in².
    b. It is 3 ½ square inches larger in area. Subtract 55 ¼ in² − 51 ¾ in² = 3 ½ in².

Fractions: Multiply and Divide Test, p. 71

1.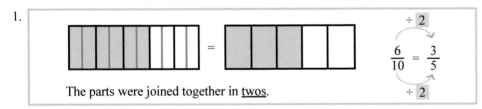
   The parts were joined together in <u>twos</u>.
   $\frac{6}{10} = \frac{3}{5}$ (÷ 2)

2. a. 3 2/3   b. 2/3   c. 1 3/32

3. She would need 5 × (2/3 C) = 3 1/3 cups of butter for 5 batches of cookies.

4. Draw five copies of 3/4: ⊕⊕⊕⊕⊕. This equals ○○○⊕ or 3 3/4.

5. It is not. The illustration shows (3/4) × (1/2) = (1/3), but in reality (3/4) × (1/2) = 3/8.

6.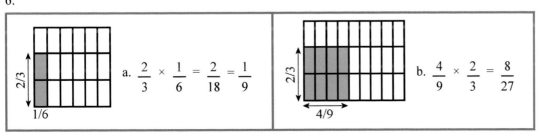
   a. $\frac{2}{3} \times \frac{1}{6} = \frac{2}{18} = \frac{1}{9}$
   b. $\frac{4}{9} \times \frac{2}{3} = \frac{8}{27}$

Fractions: Multiply and Divide Test, cont.

7. a. 5/18   b. 7 1/5

8. The area is 3 33/64 square inches.

9. Each person had 1/9 of the original pizza. (1/3) ÷ 3 = 1/9.

10. a. You can get nine servings.   b. 3 ÷ (1/3) = 9

11. a. 1/18   b. 48   c. 3/11

12. Instead of having 3 hearts, 2 circles, and 2 diamonds, a drawing could have double or triple as many of each shape. The ratio is 3:2:2.

13. One block in the model is 80. There are 160 white marbles.

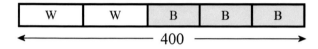

Mixed Review 11, p. 73

1. a. 2 3/6   b. 1 4/9   c. 2 5/12   d. 5 17/30

2. a. 0.28 = 2 × (1/10) + 8 × (1/100)   b. 60.068 = 6 × 10 + 0 × 1 + 0 × (1/10) + 6 × (1/100) + 8 × (1/1000)

3. 400 people. People with discounted tickets: 780 ÷ 3 = 260. Now we can find the number of people who paid the normal price by subtracting: 780 − 260 − 120 = 400.

4.

Week	Weight	Weight in ounces
0	6 lb 14 oz	110
1	6 lb 12 oz	108
2	6 lb 14 oz	110
3	7 lb	112
4	7 lb 2 oz	114
5	7 lb 4 oz	116
6	7 lb 6 oz	118
7	7 lb 7 oz	119

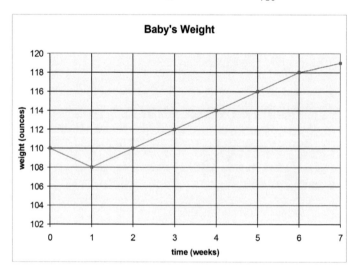

Mixed Review 11, cont.

5. **The rule for x-values:** start at 2, and add 1 each time.
   The rule for y-values: start at 0, and add 2 each time.

x	2	3	4	5	6	7
y	0	2	4	6	8	10

6. **The rule for x-values:** start at 1, and add 1 each time.
   The rule for y-values: start at 10, and subtract 2 each time.

x	1	2	3	4	5	6
y	10	8	6	4	2	0

7. Eleven. Round the price to $0.60. You can get 10 cans with $6, and one more can with the $1.

8. 4 m − 3 × 0.82 m = 1.54 m

9. a. 28 = 2 × 2 × 7   b. 55 = 5 × 11   c. 84 = 2 × 2 × 3 × 7

10. a. dog   b. Not possible, as the data is not numerical.

11. $24.96. One-tenth of $15.60 is $1.56. Two-tenths of $15.60 is therefore double that, or $3.12. Subtract that from $15.60 to find the discounted price: $15.60 − $3.12 = $12.48. Jenny bought two, so her bill was $24.96.

12. a. 3,168   b. 18,216

13. a. matches with ($170 − $23) ÷ 7 = $21. The answer tells how much John uses daily for groceries that are not treats.
    b. matches with $170 − 7 × $23 = $9. The answer tells you how much John has to use for treats.

Puzzle corner. Solutions will vary. For example:

0.4	×	0.2	= 0.08
×		×	
3	×	1.4	= 4.2
=		=	
1.2		0.28	

0.4	×	0.6	= 0.24
×		×	
0.8	×	5	= 4.0
=		=	
0.32		3.0	

Mixed Review 12, p. 76

1.

a. 13 × 4 + 18 = 70 4 + 8 ÷ 8 = 5	b. (2 + 60 ÷ 4) × 3 = 51 2 + 30 × (7 + 8) = 452	c. 10 × (9 + 18) ÷ 3 = 90 5 × (200 − 190 + 40) = 250

2. a. One hundred apples cost $23.00   b. Ten apples in each small bag is worth $2.30.

3. a. >   b. <   c. >   d. =

4.

a. 791,4<u>5</u>6,030 Place: ten thousands place Value: fifty thousand or 50,000	b. 2,0<u>9</u>4,806,391 Place: one millions place Value: four million or 4,000,000

5.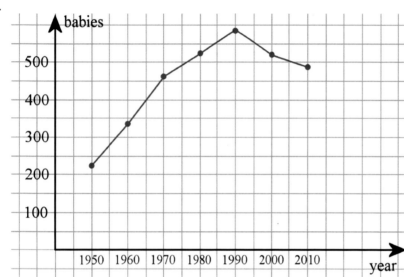

6. Equation: 4x + 176 = 516 or 516 − 176 = 4x or 516 − 4x = 176.
   Solution: x = (516 − 176) ÷ 4 = 85

7. 5 × 1.2 = 6 km

8. a. The estimated cost would be 8 × $1 + 6 × $1 = $14.
   b. Shelly's original bill is 8 × $1.19 + 6 × $0.88 = $14.80. Shelly pays 4/5 of it: $14.80 ÷ 5 × 4 = $11.84
      Or, you can multiply 0.8 × $14.80 = $11.84.

9. a. 3 8/55   b. 8 5/14   c. 6 13/20   d. 6 7/30

10. a. 1.13   b. 2.08   c. 8.94

11. a. 30 R1, 5.17   b. 10 R3, 10.75

Geometry Review, p. 79

1. a. Angles 64°, 64°, and 52°. It is an acute isosceles triangle.
   b. Angles 30°, 125°, and 25°. It is an obtuse scalene triangle.

2. a. Image on the right (not to scale):
   b. The top angle is 80°.
   c. Perimeter: 70 mm + 54 mm + 54 mm = 178 mm

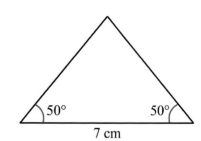

Geometry Review, cont.

3. Starting from the top left corner and going clockwise, the perimeter is:
(¾ + ½ + 2) + (1 + 1 + ½ + ½ + ½ + ½) + (2 + ½ + ½ + ½ + ¾) + 1 + 1
= 3 ¼ + 4 + 4 ¼ + 2 = 13 ½ inches.

The area is: (2 in. × ¾ in.) + (3 in. × 2 ½ in) + (½ in. × ½ in.) = 1 ½ in² + 7 ½ in² + ¼ in² = 9 ¼ in².

4. a. rhombus   b. rectangle   c. square   d. kite   e. parallelogram   f. scalene quadrilateral   g. trapezoid

5. a. rectangle   b. rhombus   c. trapezoid

6. a. a pentagon   b. c. d. There are several ways to draw the diagonals:

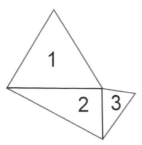 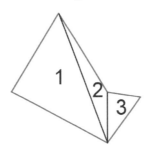 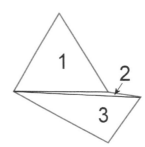

Triangle 1: Acute isosceles triangle
Triangle 2: Right scalene triangle
Triangle 3: Acute scalene triangle

Triangle 1: Acute scalene triangle
Triangle 2: Obtuse scalene triangle
Triangle 3: Acute scalene triangle

Triangle 1: Acute isosceles triangle
Triangle 2: Obtuse scalene triangle
Triangle 3: Obtuse scalene triangle

7. Answers will vary. Check students' answers. For example: a.  b.

8. a. b.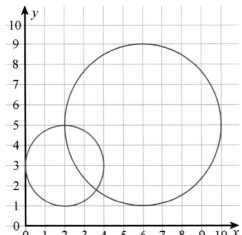

9. 4 cm. The area of the bottom is 2 cm × 4 cm = 8 cm², so the missing dimension is 32 cm³ ÷ 8 cm² = 4 cm.

10. Three boxes. One box has a volume of 6 in. × 3 in. × 2 in. = 36 cubic inches. You will need three of them to have a total volume of 108 cubic inches.

11. a. The figure has 4 × 5 × 2 = 40 cubes, so its volume would be 40 cubic inches.
    b. In this case, each little cube would have a volume of 2 in. × 2 in. × 2 in. = 8 cu. in. There are 40 cubes, so their total volume is 40 × 8 cu. in. = 320 cu. in. Or, you can calculate the volume by first calculating the three dimensions: the length is 10 inches, the depth is 4 inches, and the height is 8 inches, so the volume is then 10 in. × 4 in. × 8 in. = 320 cu. in.

Puzzle corner. The edge length of the cube must be 4 cm. Therefore, its volume is 4 cm × 4 cm × 4 cm = 64 cm³.

Geometry Test, p. 83

1. a. rhombus b. trapezoid c. right scalene triangle d. isosceles obtuse triangle e. kite

2. Yes, it is a kite, because it has four congruent sides it is a rhombus, and all rhombi are also kites.
   Yes, it is a trapezoid, because all parallelograms are trapezoids. It may or may not be a square (we do not have enough information).

3. a. Yes, it is. A kite has two pairs of congruent sides, and the congruent sides are adjacent. In a square, all sides are congruent, so that fulfills the definition of a kite.

   b. Yes, it is, because it has two sides that are parallel.

   c. Yes, it can, if it is a square. Check the student's sketches; they should show a square.

   d. No, it cannot. An equilateral triangle has three angles that are 60°; therefore it does not have any right angles.

4. a. perimeter b. volume c. area

5. Students' triangles will vary, but should have the same basic shape as the example triangle on the right. Start out by drawing the base side. Then draw the 30°-angles and continue the sides until they meet. The top angle measures 120°.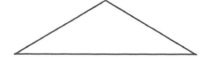

6. The area is 1 square inch.

7. The volume is 2 ft × 1.5 ft × 1.5 ft = 4.5 cubic feet.

8. a. The volume of one book is 15 cm × 30 cm × 1.5 cm = 675 cm³.
   b. The volume of six books is 6 × 675 cm³ = 4,050 cm³.

9. The triangle is acute and scalene.

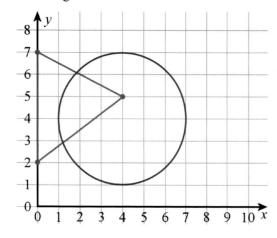

Mixed Review 13, p. 85

1. First, subtract $80 − $12 = $68. Then take half of that: $34. Angela got $34 and Eric got $46.

   Angela [$34]
   Eric [$34][$12] } $80

2. $6. One orchid costs $8, and one daisy costs $2. The price difference is $6.

3. a. 215. Check: 46 × 215 = 9,890
   b. 1.1 Check: 65 × 1.1 = 71.5

4. a. 0.75 b. 0.64 c. 0.09 d. 7.2 e. 0.01 f. 4.2
   g. 0.2   h. 8   i. 0.9   j. 2   k. 200   l. 0.008

5.

a. $\frac{6}{5} = 1\frac{1}{5}$	b. $\frac{2}{21}$	c. $\frac{3}{44}$
d. $\frac{5}{9}$	e. 15	f. 15
g. $\frac{1}{10}$	h. $\frac{1}{30}$	i. $\frac{7}{5} = 1\frac{2}{5}$
j. $\frac{4}{9}$	k. $\frac{40}{3} = 13\frac{1}{3}$	l. $\frac{62}{9} = 6\frac{8}{9}$

6. Four jars: 4 × (1 3/8 in.) = 4 12/8 in. = 5 1/2 in.

7. a. x = 4 × 14 = 56   b. x = 1,500 ÷ 6 = 250

Mixed Review 13, cont.

8. See the recipe on the right.

   What do you think she should do with the eggs?

   She should use 1 egg. Technically, 2/3 of 2 eggs is 1 1/3 eggs, but that is not practical to use.

   Pancakes
   2 2/3 dl water
   1 egg
   2 dl whole wheat flour
   (pinch of salt)
   33 g butter for frying

9.

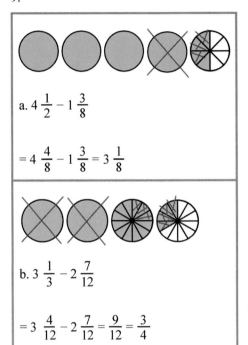

a. $4\frac{1}{2} - 1\frac{3}{8}$

$= 4\frac{4}{8} - 1\frac{3}{8} = 3\frac{1}{8}$

b. $3\frac{1}{3} - 2\frac{7}{12}$

$= 3\frac{4}{12} - 2\frac{7}{12} = \frac{9}{12} = \frac{3}{4}$

10. a. 3:8   b. 5:8   c. 5:3

11. a.

Museum's visitors			
Day	Adults	Children	Total Visitors
Monday	29	14	43
Tuesday	23	10	33
Wednesday	34	18	52
Thursday	38	19	57
Friday	35	19	54
Saturday	57	25	82
Sunday	63	31	94
Totals	279	136	415

b. Sunday is the busiest day with 94 visitors, and Tuesday the least busy with 33 visitors. The difference in the total visitor count between those two days is 94 − 33 = 61.

c. $279 \div 7 = 39.9$

d. $136 \div 7 = 19.4$

e.
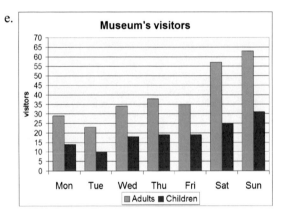

Mixed Review 14, p. 88

1.

Picture or Diagram	As Fractions	As a Ratio
	1/6 of the shapes are hearts. 5/6 of the shapes are diamonds.	The ratio of hearts to diamonds is 1:5.

2. a. 1 : 3   b. 3 : 6

3. 24 were oatmeal cookies.

4. 33 cubic inches

5. The estimates may vary.
   a. Estimate: 22 × 4 = 88   Exact: 84.63
   b. Estimate: 0.5 × 1 = 0.5   Exact: 0.416
   c. Estimate: 140 × 5 = 700   Exact: 736.02

6. a. 91.5 km   b. 40 × 5 × 91.5 km = 18,300 km

7. a. kite   b. rectangle   c. trapezoid   d. parallelogram

8. The only way to draw this is if the two sides that are 6 cm long are the ones that meet in a right angle. (The image is not to scale.)

9. a. 1/6   b. 1 10/11

10. You can fill ten glasses.

11. a.

AGE (yrs)	WEIGHT (kg)	Weight gain from previous year	AGE (yrs)	WEIGHT (kg)	Weight gain from previous year
0	3.3 kg	-	10	31.4 kg	3.3 kg
1	10.2 kg	6.9 kg	11	32.4 kg	1.0 kg
2	12.3 kg	2.1 kg	12	37.0 kg	4.6 kg
3	14.6 kg	2.3 kg	13	40.9 kg	3.9 kg
4	16.7 kg	2.1 kg	14	47.0 kg	6.1 kg
5	18.7 kg	2.0 kg	15	52.6 kg	5.6 kg
6	20.7 kg	2.0 kg	16	58.0 kg	5.4 kg
7	22.9 kg	2.2 kg	17	62.7 kg	4.7 kg
8	25.3 kg	2.4 kg	18	65.0 kg	2.3 kg
9	28.1 kg	2.8 kg			

b. He gained the fastest at ages 1 year, 14 years, 15 years, and 16 years.
c. You can see that by how steeply the line is rising on the graph.

End-of-the-Year Test, p. 91

Please see the file for the End of the Year Test for grading instructions.

The Four Operations

1. a. 45   b. 409,344

2. a. $x = 296{,}430$ b. $Y = 80$ c. $N = 3{,}304$

3. All of these are correct:
   $4Y = 600$ or $4 \times Y = 600$ or $Y + Y + Y + Y = 600$ or $600 \div 4 = Y$ or $600 \div Y = 4$ or $600 - Y - Y - Y - Y = 0$.
   Solution: $Y = 150$.

4. a. $42 \times 10 = (10 - 4) \times 70$ b. $143 = 13 \times (5 + 6)$

5. $(\$19.95 - \$5) \times 5$ or $5 \times (\$19.95 - \$5)$. Her total cost was $74.75.

6. No, it is not. Explanations will vary. For example: It is an odd number, and therefore cannot be divisible by an even number. $991 \div 4 = 247$ R3, leaving a remainder, so 991 is not divisible by 4.

7. a. $26 = 2 \times 13$ b. $40 = 2 \times 2 \times 2 \times 5$ c. 59 is prime

Large Numbers

8. a. 70,016,090   b. 32,000,232,000

9. It is about $32{,}000 \times 300 = 9{,}600{,}000$. Other estimates are also possible.

10. 80 million or 80,000,000

11.

number	593,204	19,054,947
to the nearest 1,000	593,000	19,055,000
to the nearest 10,000	590,000	19,050,000
to the nearest 100,000	600,000	19,100,000
to the nearest million	1,000,000	19,000,000

Problem Solving

12. An 8-ft long board is 96 inches. One-sixth of that is 96 in. $\div$ 6 = 16 in. The remaining piece is 80 inches, or 6 ft 8 in.

13. It would cost $7.80 to download ten songs. First, find the price of one song download: $4.68 \div 6 = \$0.78$. Then, multiply that by 10.

14. A lunch in the cheap restaurant costs 1/3 of $36, or $12. Mary spends $36 + 4 \times \$12 = \$84$.

15.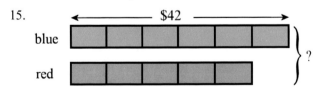

    One block in the model is $42 \div 6 = \$7$. The red swimsuit costs $5 \times \$7 = \$35$. Together they cost $77.

16. a.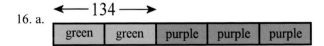

    b. One block or part in the model is $134 \div 2 = 67$ marbles. There are $3 \times 67 = 201$ purple marbles.

17. a. The DVD costs about $30. Karen pays 3/5 of it, which is about $30 \div 5 \times 3 = \$18$. Ann pays about $12.
    b. Karen pays $29.90 \div 5 \times 3 = \$17.94$. Ann pays $11.96.

End-of-the-Year Test, cont.

Decimals

18. a. 0.289 b. 0.30 c. 0.305 d. 0.313

19. a. 0.95 b. 0.72 c. 0.62 d. 1.26 e. 1.05 f. 0.37

20. a. 0.08 b. 0.081 c. 5.21

21. a. $\frac{48}{1000}$ b. $1\frac{4}{1000}$ c. $7\frac{22}{100}$

22. a. 0.31 > 0.031   b. 0.43 > 0.093   c. 1.6 > 1.29

23.

rounded to...	nearest one	nearest tenth	nearest hundredth	rounded to...	nearest one	nearest tenth	nearest hundredth
5.098	5	5.1	5.10	0.306	0	0.3	0.31

24.

a. 0.4 × 7 = 2.8 b. 0.4 × 0.7 = 0.28 c. 0.4 × 700 = 280	d. 10 × 0.05 = 0.5 e. 100 × 0.05 = 5 f. 1000 × 0.5 = 500	g. 1.1 × 0.3 = 0.33 h. 70 × 0.9 = 63 i. 20 × 0.09 = 0.18

25.

a. 0.36 ÷ 6 = 0.06 b. 5.6 ÷ 7 = 0.8	c. 3 ÷ 100 = 0.03 d. 0.7 ÷ 10 = 0.07	e. 16 ÷ 10 = 1.6 f. 71 ÷ 100 = 0.71

26.

a. 0.2 m = 20 cm 37 cm = 0.37 m 2.9 km = 2,900 m	b. 0.4 L = 400 ml 3.5 kg = 3,500 g 240 g = 0.24 kg	c. 56 oz = 3 lb 8 oz 74 in. = 6 ft 2 in. 15 C = 3 qt 3 C

27. There are 444 milliliters in two bowls. Two liters is 2,000 ml. 2,000 ml ÷ 9 = 222.2 ml or about 222 ml.

28. a. 1.42   b. 14.28   b. 14.08

End-of-the-Year Test, cont.

Graphs

29.

x	0	1	2	3	4	5
y	1	3	5	7	9	11

30. See the image on the right.

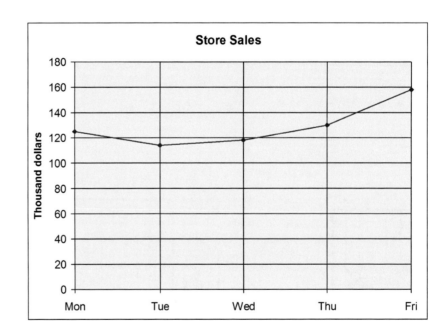

31.

Day	Sales (1000 dollars)
Mon	125
Tue	114
Wed	118
Thu	130
Fri	158

a. See the line graph on the right.

b. The average daily sales is $129,000.

133

End-of-the-Year Test, cont.

Fractions

32. a. 6 1/3   b. 2 1/3   c. 13 4/5

33.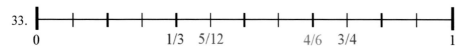

34. | a. ~~5/6 = 5/20~~ | b. $\frac{2}{7} = \frac{8}{28}$ | c. $\frac{3}{8} = \frac{15}{40}$ | d. $\frac{2}{9} = \frac{6}{27}$ |

35. Mia found the common denominator (15) correctly, but forgot that the 2 fifths and the 2 thirds do not stay as 2 fifteenths in the conversion.

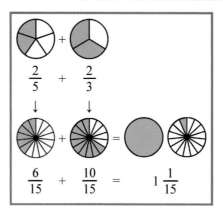

36. 1 1/6   b. 7/15   c. 5 5/8   d. 10 5/18

37. You would need 3 × (2 3/4) = 8 1/4 cups of flour to make three batches of rolls.

38. a. $\frac{6}{9} > \frac{6}{13}$ b. $\frac{6}{13} < \frac{1}{2}$ c. $\frac{5}{10} > \frac{48}{100}$ d. $\frac{1}{4} = \frac{25}{100}$ e. $\frac{5}{7} > \frac{7}{10}$

39. a. 1 2/5   b. cannot be simplified   c. 7/8

40. Yes, it is correct. (2/3) × (1/2) = 1/3.

41.

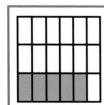 a. $\frac{1}{3} \times \frac{5}{6} = \frac{5}{18}$ 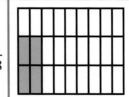 b. $\frac{2}{9} \times \frac{2}{3} = \frac{4}{27}$

42. You can cut 60 pieces. 15 in. ÷ (1/4 in.) = 60

43. 1/6 of the pizza. (1/2) ÷ 3 = 1/6

44. a. 10 1/2   b. 1/21   c. 2 14/15   d. 18

End-of-the-Year Test, cont.

Geometry

45. The sides measure 2 15/16 in., 2 9/16 in., and 4 15/16 in. The perimeter is 10 7/16 in.

46. a. an isosceles acute triangle   b. a rhombus   c. a right scalene triangle   d. a trapezoid

47. a. 9 m^2 b. 20 ft

48. Yes, it is. A square has one pair of parallel sides, which is a definition of a trapezoid.

49. Yes, it can. For example 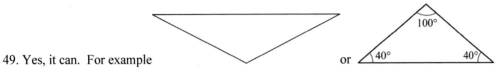 or

50. a. Check the student's triangles. The student should use a tool, such as a triangular ruler or a protractor, to make the right angle. The picture below may be slightly out of scale when printed, due to the possible scaling in the printing process.

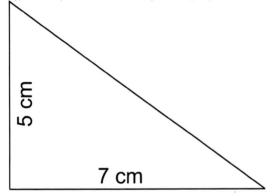

b. 8.6 cm + 5 cm + 7 cm = 20.6 cm
c. They measure 90 °, 36 °, and 54 °.

51. The volume is 5 cm × 10 cm × 4 cm = 200 cm^3.

52. a. 1.2 m × 0.6 m × 1 m = 0.72 m^3.
    b. 240 liters. 0.72 m^3 is 720 liters, and one-third of that is 240 liters.

Made in the USA
Middletown, DE
08 August 2017